D1567720

TALKING BACK
to the MACHINE

PETER J. DENNING
Editor

TALKING BACK to the MACHINE

COMPUTERS and HUMAN ASPIRATION

Introductions by
JAMES BURKE

COPERNICUS
AN IMPRINT OF SPRINGER-VERLAG

Except for Peter Denning's introduction and David Kasik's essay, the chapters of this book are speeches presented at the ACM97 conference in San Jose, California, during the first week of March 1997. With the authors' permission, they have been edited extensively to convert them into literary form. Every effort was made to preserve the spirit of the spoken word so that each essay reflects not only the author's voice, but the author's meaning.

—*The Editors*

Published in the United States by Copernicus, an imprint of Springer-Verlag New York, Inc.

Copernicus
Springer-Verlag New York, Inc.
175 Fifth Avenue
New York, NY 10010
USA

Library of Congress Cataloging-in-Publication Data
Denning, Peter J., 1942–
Talking back to the machine : computers and human aspiration / Peter Denning.
p. cm.
Includes bibliographical references and index.
ISBN 0-387-98413-5 (hardcover : alk. paper)
1. Computers and civilization. 2. Prediction theory. I. Title.
QA76.9.C66D46 1998
303.48′34—dc21 98-41896

Manufactured in the United States of America.
Printed on acid-free paper.

9 8 7 6 5 4 3 2 1

ISBN 0-387-98413-5 SPIN 10663208

To our grandchildren,
who will live with the consequences of what we say
here and who will probably invent something
better than anything we have imagined.

CONTENTS

PETER J. DENNING

Then Now

The world's electronic computers, first switched on a little more than 50 years ago, have wrought enormous changes in how we live, work and see ourselves. In 1947, at the dawn of the computer age, a few far-sighted visionaries founded ACM, the Association for Computing Machinery, to help nurture the new field and the people in it. A half century later, the ACM celebrated its own golden jubilee with ACM97—a conference, an exposition, a Web site, and a book, *Beyond Calculation: The Next Fifty Years of Computing* (Copernicus). In what *The New York Times* called "essays of astonishing intellectual reach," 24 leading thinkers in the industry described the future of computing as they envisioned it unfolding. Writing for the general reader, they wove thought-provoking essays of remarkable insight and depth. They discussed everything from new technological developments to how computing may affect children, workplace styles, education, research and business innovation.

The conference itself was so stimulating, and the interest in the first book so great, that we produced a sequel entitled, *Talking Back to the Machine: Computers and Human Aspiration*. It features discussions of how computers will influence how we live, learn, teach and communicate with each other in the coming decades. Essayists include the Nobel Prize winner Murray Gell-Mann on information quality, former Secretary of Defense William Perry on computers in war, computer pioneer Maurice Wilkes on surprises, international business leader Fernando Flores on business communication, Walt Disney Vice President Bran Ferren on storytelling with new media, academic leader Elliot Soloway on education and TV producer James Burke on connections between people and technologies. These and the other writers describe myriad ways, both good and bad, in which our lives will be altered by information technology and how we might influence the shape of things to come.

Charles Babbage, a visionary mechanical engineer, conceived the first general-purpose computing engine about 150 years ago. He built only part of it. Almost a hundred years passed before electrical engineers teamed with mathematicians to build automatic calculating machines. Electronics and a theory of algorithms succeeded where pure mechanics did not. The first electronic computers, built as part of the 1940s war effort, were intended for scientific and mathematical calculation: They figured ballistic trajectories and broke ciphers. Many alumni of these projects started computer companies. They saw the main opportunities for growth in the computer's potential for driving international business and for intelligent actions: IBM and Univac bet on business data processing while the newspapers contemplated the implications of "electronic brains." Thomas Watson, the founder of IBM, opined that at most four or five computers would be needed for scientific calculations worldwide.

Others soon joined IBM and Univac in the computer marketplace, adding names like RCA, Burroughs, ICL, General Electric and Control Data to the growing cadre of computer companies. (Many of those early companies have since disappeared or left the computer business.) People working in these companies and research labs have produced an amazing array of innovations over the past 50 years. They designed the first programming languages—FORTRAN, ALGOL, COBOL and LISP—and the first operating systems in the mid-1950s. They formed the first computer science departments in universities in the early 1960s. In 1972, Hewlett-Packard's hand calculator made the slide rule obsolete. The first computer chips appeared in the late 1970s along with a plethora of personal computers aimed mostly at hobbyists.

In 1981, IBM transformed the personal computer into a business; the IBM PC became an industry standard, the machine to be imitated (as indeed it was by many clone-makers). Bob Metcalfe invented the Ethernet in 1973 at Xerox, allowing many computers to talk over a single coaxial cable. During the next ten years he transformed the Ethernet into an industry standard for local networks connecting PCs. By 1990, word processing, spreadsheets, computer-aided design systems and database programs respectively made the clerk typist, clerk accountant, draftsman and record-keeper obsolete. The Internet, which descended from the ARPANET of the 1970s, and the World Wide Web, which seemed to appear out of nowhere in 1992, propelled personal computers—linked by local networks and modems—to the center of international business practice. As computers shrank, mainframes became obsolete—except as computing engines for large applications in business, science and engineering.

Over those 50 years, the cost and size of computers dropped steadily: Today, $2,000 will buy you a computer that is 1,000 times faster and holds 1,000 times as much data as the $1 million mainframes of the 1950s. The computer revolution has pervaded the lives of vast numbers of people: By the beginning of 1999, there were over 80 million users on 30 million computers serving up more than 400 million Web pages worldwide on the Internet.

In the midst of this chaos and ferment, astute observers have spotted trends of remarkable stability. One of the most famous patterns is Moore's law, an empirical observation of Gordon Moore, one of the founders of Intel. Moore said that the speed of microprocessors doubles every 18 months. Since the start of the computing age, there have already been 18 such doublings; experts now argue about how many more we can expect before the physical limits of miniaturization halt the process. Similar trends have been identified in the growth of the Internet and in the capacities of data networks to transfer files and images. Extrapolate these trends from curiosities now gestating in research labs, and we have what looks like a solid basis for predicting what will happen with computing in the decades ahead. Or do we?

Fifty Years of Surprises

History teaches us a different lesson. Minicomputer pioneer Gordon Bell and Maurice Wilkes remind us that confidence in long-range predictions may be hubris. Although we have been able to predict certain events by extrapolating trends over a short term, we have been notoriously poor at predicting what people will do with any technology in the long term. What people *do* is part of human practices, which stoutly resist quantitative analysis.

Imagine that Henry Ford could return to see today's automobiles. He would hardly be surprised by the changes in design: Cars still have four wheels, steering, front-mounted internal-combustion engines, transmissions and the like. But he would be greatly surprised by the changes in human practices that have grown up around the automobile—for example, hot rods, strip malls, drive-in fast-food chains, rush hours, traffic reports, stereo systems, mobile phones, navigator systems, cruise controls and more. Alexander Graham Bell would similarly be little surprised by the design of telephone systems, but practices like "prestige" exchanges, telemarketing and telephone pornography would astound him. Can you imagine trying to explain lava lamps to Edison or frequent-flyer miles to the Wright brothers?

What has happened with computing and telecommunications has certainly been a surprise for me. I was born in the early days of the current revolution. I have been interested in science since childhood; astronomy and botany since grammar school; electronics since middle school; and computers since high school. As a graduate student in the 1960s, I was immersed in the MIT optimism about the possibilities offered by computing technology. We were optimistic that one day computers would shrink to fit on top of a desk (or, at least, *be* the desk). We were optimistic about far-flung networks and resource sharing, about graphics and about artificial intelligence. Deep down, however, all this seemed like wishful thinking. For if you told me that the slide rule would be obsolete within five years of my graduation, the typewriter within ten years, or the publisher-owned copyright of research papers within 30; or that Internet addresses (whatever they were) would be displayed on business cards and television ads, that people would give up their home telephones for cell phones; or that new computers would be designed based on DNA, nanotechnology, quantum mechanics or biological silicon, I would have thought you were crazy. But here we are, with all this and more. I am grateful to have lived to see my romance with computing technology be requited. Go ahead and dream—some of them are likely to come true!

Today Foreseen

In our charge to the authors of *Beyond Calculation* and the speakers at ACM97 to look ahead, Bob Metcalfe and I counseled against baseless predictions. New-millennia predictions are as plentiful (and as cheap) as grains of sand. In response, most of the authors and speakers did one of two things: They looked at trends or at human nature, figuring that they could extrapolate the trends or count on human nature being the same.

Despite all the protestations of the editors and authors, our readers and listeners heard predictions and pondered their accuracy. The speculations are, after all, advanced by the industry's great thinkers. Surely what they say is more likely to come to pass than what others might say.

Fortuitously, an event 100 years ago offers guidance on this question. In 1893, the Fourth World's Columbian Exposition was held in Chicago. It celebrated the 400th anniversary of Columbus's landing in the Western Hemisphere. Like other world's fairs, it sought to demonstrate future possibilities in science, technology, art and culture. It also featured a look ahead to the Fifth Columbian Exposition in 1993. (Curiously, the 1993 event never took place.)

The American Press Association organized a group of 74 leading authors, journalists, industrialists, business leaders, engineers, social critics, lawyers, politicians, religious leaders and other luminaries of the day to give their forecasts of the world 100 years later. Their 74 commentaries were published in the national newspapers for several months preceding the Exposition. One hundred years later journalist and historian Dave Walter compiled and republished their commentaries in a volume he called *Today Then: America's best minds look 100 years into the future on the occasion of the 1893 World's Columbian Exposition* [American World & Geographic Publishing, 1992]. In reading these old essays, we learn more about the writers and how they observed their world than we do about our own world.

Among the most striking features of the 1893 forecasts is the remarkable paucity of predictions that actually came true. Some of them seem outlandish, completely disconnected from reality—but fervently believed by their authors. For example, religious leader Thomas De Witt Talmage thought that longevity would be increased to 150 years. U.S. Senator W.A. Peffer thought that pollution would no longer be a problem. Comptroller of the U.S. Treasury Asa Mathews thought the USA would include Canada and Mexico in a total of 60 states (there were 44 then). Editor and publisher Erastus Wiman thought that there would be minimal taxation, worldwide free trade and no standing army. Engineer George Westinghouse thought that trains would operate at speeds of 40–60 miles per hour and that faster speeds, though possible, were too unsafe. Railroad icon and lawyer T.V. Powderly thought there would be no very rich or very poor, that no family would have more children than it could sustain and that divorces would be rare. Commissioner of Indian Affairs Thomas Morgan thought that Indian tribes would disappear and be replaced with a highly respected Indian cultural tradition. Lawyer and politician William Jennings Bryan predicted the abolition of the electoral college in presidential elections.

Many thought railways would be the primary method of transportation, extending from the northernmost parts of Canada to the southernmost parts of South America. They thought that pneumatic tubes would be common modes of transportation for people in cities as well as a means of moving mail transcontinentally. They thought government would be smaller and that there would be fewer class differences. Few foresaw the world wars, the communications revolution or air transportation. None foresaw the interstate highway system, genetic engineering, mass state-sponsored education or broadcast TV and radio—or the computer.

These commentators, probably reflecting more widely held opinions of the day, were particularly possessed by two beliefs: that technology would solve society's ills, and that people would change dramatically for the better. Some spoke as if the changes they forecast were inevitable; some simply prayed for solutions to social problems; some attempted to extrapolate trends. The few commentators who came closest to describing the world as we know it today were the most skeptical about the idea of technology solving our problems and about the mutability of human nature.

Animated by Our Beliefs

What can we learn about our own world by reading the forecasts of our ACM97 authors and speakers? What can we learn about *now* by reading about *then?* I discovered six unspoken presuppositions running through many of the essays.

(1) *Technology will continue to progress at an ever-increasing rate, producing generally positive changes.* We believe in extrapolations like Moore's law, the diminishing significance of distance, the flattening of communication costs. We accept the motto "change is the only constant." We believe that Moore's law runs out, when new technologies will be available to continue or accelerate the rate of change, bringing benefits faster. The possibility that some outcomes may be negative is discussed but not taken seriously—for example, that cost-efficient national medical databanks formed by health management organizations may trample individual privacy rights, that detailed surveillance to deter intruders might enable control-oriented managers to make the workplace distinctly unpleasant, or that government's power to protect citizens might be eroded by its inability to collect taxes.

Computing research visionary Joel Birnbaum and semiconductor innovator Carver Mead explicitly discuss new technologies that will enable the exponential growth of computing power to continue. Internet pioneer Vint Cerf and former FCC Chairman Reed Hundt are certain that communication bandwidth will become inexorably cheaper and wider. Robotics pioneer Raj

Reddy sees a time when such improvements will permit virtual time travel, virtual teleportation and immortality for those willing to survive as disembodied intelligences in cyberspace.

(2) *Technology drives social and commercial change, placing technologists in a special stewardship.* Economists want us to believe that prosperity has resulted from policies they have been able to recommend from their simulations of economies. Entertainment companies want us to believe that their storytelling abilities and production of worthy content have saved the Internet from being a barren wasteland. Business leaders want us to believe that their spending decisions, driven by customer needs, determine which technologies and services can actually be supported. Baseball players would have us believe (at least in 1998) that the feats of home run hitters Mark McGwire and Sammy Sosa, brought to us by the national networks and backed by sophisticated databases, have elevated the national mood—and with it prosperity. We technologists are no different. We would have others believe that the advances we produce drive all the other changes they cherish. The scenario of the Year 2000 Date Bug bringing on the collapse of civilization strikes us as amusing but unlikely because, after all, *we* will summon the brainpower and the technology itself to overcome the problem at the last minute and keep cyberspace humming. Our technology is supreme!

The truth is that all these factors, and more, play together in an intricately complex game whose evolution we call progress. The possibility that other players and forces might affect change more than our technologies may not appeal to many of us. The possibility that we do control the direction of the technology frightens some of us.

Notable among the dissenters from this view are Elliot Soloway, who sees the education of children as a fundamentally human activity; and science fiction writer Bruce Sterling, who believes that the dark side of human nature will express itself at the slightest chance.

(3) *Surprises will abound.* Who hasn't seen the list of offbase predictions such as Bill Gates's claim that no one would ever need more than 640K of memory? Or the claims in the original ARPANET documents that resource-sharing rather then E-mail would be the driving force in networking? Or the derisive critiques claiming that 1960s software engineers mistakenly assumed that the COBOL language and database formats would be long dead by the year 2000? These statements might represent surprises, but none is of the magnitude of the sweeping changes that few of the 1893 prognosticators foresaw. We speak of the folly of prediction and then give forecasts with the conviction of an astronomer pinpointing the time of tomorrow's sunrise.

If we really believe the rhetoric about surprises, why don't we look more systematically from whence they came? Technological surprises (breakthroughs) most often come from the "boundaries"—interactions between people of different domains exploring a common interest. Business surprises come from marginal practices—those at the boundaries of a field—that can be broadened to solve a major problem or produce an enormous benefit. (The World Wide Web, which started out as a means for physicists to exchange research papers, was like this.) The more remote the boundary, the bigger the surprise.

Entrepreneurs are more familiar with the phenomenon of boundaries than most researchers. The number of boundaries between information technologists and other domains grows as computers invade ever more diverse fields. Today's boundaries include: biology, notably DNA computing, organic memories, bionic body parts and sensors, 3D real-time imaging; physics, including materials, photonics, quantum computing; massive Internet computations; neuroscience, cognitive science, psychology; large-scale models for climate, economics, aircraft simulation, earthquake prediction and weather forecasting; data mining from massive data sets; library sciences; workflow and coordination in organizations; humanities, arts, music and storytelling. Researchers give lip service to these boundaries but relatively few embrace them passionately. No wonder so many are surprised when an invader from one of the frontiers crashes through our quiet neighborhoods.

(4) *Computers can—and should—be a leveling force, eliminating class differences and pulling up the indigent.* There is much talk on this topic: Information wants to be free; no government can successfully restrict the flow of information (or funds) across its national boundaries; computers offer instant democracy; computers can make education universal; and an individual's personal power increases through access to knowledge via the Internet. A better age is coming (but some help from government is needed in the short term to prevent the gap between the haves and have-nots from widening).

This proposition is none other than the two-century-old belief that technology is capable of solving many of society's problems. Yet the same problems are with us today. This proposition relieves us of personal responsibility to help others—after all, the technology will do it if we just give it time. It deserves our undying scorn.

(5) *Computers and information are great metaphors for understanding how things work.* Computers have given us new ways of thinking about machines, communications, organizations, societies, countries and economies. Throughout the ages, every technology has given us metaphors for nature

and ourselves. René Descartes's mind–body dualism, for instance, proposed that a spirit inhabited an extraordinarily complicated clock-like device—the "ghost in the machine." Freud's theory of the unconscious implicitly evoked the image of a steam engine: Impulses blocked from their natural release would build up pressure in the subconscious mind until they blew out elsewhere, often far from their point of origin. Today, neuroscientists routinely talk about "feedback loops" and "brain circuitry"; instinctive behaviors are said to be "wired" or "programmed"; we possess "software" for certain kinds of mental activity. Educators describe learning as a process of transferring information from a corpus of knowledge to the student.

It is respectable in the 1990s to talk about all mental activities, including visual imagery and memory recall, as algorithmic processes. We have spent a substantial part of the past half century trying to build computers that resemble the mind—all we have to show for it are minds that think they resemble computers. Computer scientists have not been satisfied to automate our ability to calculate; they have attacked object recognition, language comprehension, reasoning about the world, digitizing the world's libraries, and much more. As the steam engine was an iron horse, the computer is a silicon brain.

Cracks are already beginning to appear in the computing metaphor. A growing number of educators, for example, say that there is much more to learning than transferring information; they say the phenomenon of embodied knowledge, learned through practice and involvement with other people, is a process that cannot be understood simply as information transfer. Terms from biology and genetic engineering are beginning to creep in; for example, more economists describe economies as ecologies rather than as engines of growth.

Who ventures guesses about the great metaphors 50 years hence? Fernando Flores defines the age of identities for the business world, and James Burke defines an age of connections for society.

(6) *Virtual reality blurs the distinction between what is real and what is not.* Some computer scientists have become enamored of virtual reality—the full-scale simulation of all sensory input that a person can experience such that the person cannot easily distinguish the simulated world from the real world. Virtual reality can help people learn important skills in settings where mistakes don't have real consequences. It is promoted as the courier of great benefits, allowing you to "walk" around in your new house or office before committing to construction, teaching pilots how an aircraft feels before they actually fly it, helping people learn French by immersing themselves in a simulated France, training software managers by having them manage a simulated project, and much more. These benefits are real.

Yet it is easy to get carried away with these speculations to the point where you doubt that anyone will care about the everyday world since they will be able to "jack in" to whatever reality interests them at the moment. What we call reality begins to look more and more like a social construction and does not need to be grounded in real phenomena. Extra-scientific phenomena such as telepathy, telekinetics, time transport, hyperspace flight, wormholes and artificial life are given equal footing with established, well-grounded science. These beliefs aren't restricted to naive viewers of modern science fiction movies—some of the ACM97 conference speakers embrace them, like software guru Nathan Myhrvold telling his audience (one hopes with tongue in cheek) that the essence of his personality and the genetic code that distinguishes him from anyone else can be captured on a 1.44 MB floppy disk and that his next life will be spent as a virtual personality roaming the Internet.

Powers of Imagination

Some have said that the great science-fiction writers have been right more often than our leading thinkers. They base their claim on science-fiction stories written fifty years ago that describe scenes familiar today. But, like the six propositions above, this claim is easier to believe than to verify. I've never seen any data analyzing the science fiction literature. I know there is a huge amount of "pulp science fiction"—throwaway stuff that isn't worth reading. So what if one in 50,000 stories contains elements of truth about today? That doesn't tell us much about the predictive abilities of science fiction writers or how to find the ones who make sound forecasts.

The importance of the essays in this book does not depend on their value as science fiction. They reinforce something we already know about leadership: Leaders with powerful stories that inspire the imagination and generate worthwhile possibilities for people are the ones who inspire followers to make their dreams come true. Our authors are all industry leaders. The power of their imagination will draw people into the worlds they see.

JAMES BURKE

Opening Connections

Introduction

Good morning ladies and gentlemen. This conference is going to look at the next 50 years of computing and, in doing so, it will examine the process of change in a field that has already generated more changes in the last half century than perhaps in the whole of history that came before. And that rate of change is accelerating. In the world of ordinary users, where I come from, this often leaves people confused. By the time you get around to reading the manual these days, the model is already obsolete—if you can understand the manual in the first place.

This situation always reminds me of the story of the depressive who gets a few days off from the hospital and goes to the beach to get himself a tan. A couple of days later, back at the hospital, his psychiatrist gets a postcard from his patient. The message reads, "Having a wonderful time. Why?" The problem, of course, is that because of the serendipitous way in which change itself happens, and the number of wildly different inputs that can be involved, it is often difficult to be ready for change by second guessing it.

Take the case of the toilet roll. We know that when the Chinese invented paper, this probably was not the first use they had in mind for it. No, the history of paper is not based on the ineluctable move over the centuries toward the bathroom. The continuous-roll paper-production system that made the toilet roll possible arrived in England in the mid-19th century, just when the cholera epidemic, new sewage systems and vitreous china lavatories put the idea of hygiene into people's heads. But the continuous-roll process itself only got there in the first place because some guys accidentally picked up the patent for it in France while buying the patent for something else: a food-preservation system invented by a champagne bottler in France to feed Napoleon's armies. Those armies were winning their wars with highly mobile horse artillery thanks to lightweight cannons made from a new cannon-boring machine invented by an Englishman using a new kind of steel for the cutting head. The steel was invented a little bit earlier in order to provide a better clock spring for navigators going west to exploit the colonists. They needed to know what time it was back at base. Now the development of the new navigational gizmo that made use of that spring—called Harrison's chronometer—was triggered in the first place by a gigantic prize offered by the British Parliament after the ill-fated journey of an admiral named Sir Cloudsly Shovel. His fleet, which was on its way back to England one foggy night in 1707, made a serious navigational boo-boo and hit some rocks. The whole fleet sank and all his men drowned. So it could be said that the toilet roll comes from serious 18th century navigational problems. But look at the series of events toward the emergence of this humble object: navigation technology, clocks, steel, artillery, food preservation, disease, vitreous china and sewage systems. See what I mean about it being hard to second guess the process of change?

The other thing that gives this conference particular relevance has to do with the subject of the conference itself: Computers. Because of information technology, we stand today on the threshold of a social and scientific revolution the likes of which will make everything that came before look like slow motion. What makes it all the more challenging to forecast what that revolution will bring is the fact that all advances in information technology through history seem to cause an explosion of innovation. That, in turn, causes unforeseen secondary effects that ripple out in all directions.

Take the alphabet. It seems to pop up in a place called Serabit el Khadem down on the Sinai Peninsula around 1500 B.C., primarily as a system for unifying the various forms of hieroglyphic gobbledygook so that the Phoenician mining contractors could do easier deals with everybody. That, according to some scholars, made it possible for the Greeks to use alphabetic thinking to develop logic and the reductionist thought processes that break the universe down into its smallest components. Eventually, this gave birth to the specialist gobbledygook now spoken by any Ph.D. whose subject is not your own and the concomitant modern problem of change generated by people whose prime aim is to know more and more about less and less.

A pal of mine at Oxford, for example, got his doctorate on Milton's use of a comma and 17th century printed maps. What the maps did was encourage voyages of exploration that needed all kinds of things to be invented for that purpose. The result: new land registers to raise backers' money, with new mortgage companies to facilitate that, new joint stock companies to take the risk out of investment and, through them, a stock market to trade those stocks on. This was run by a new national bank, which provided financing through new credit agencies for new enterprises running on another new thing called a business contract whose mutual-obligation clause was inspired by the Constitution of another new thing called the United States of America.

So there seems to be two key patterns to be aware of: the sometimes marginal areas in which the domino effect of one field of innovation can cause major effects in another; and the powerful ripple effect spreading out around each advance in the field of information and communications technology. A third element in the process by which change alters life seems to be in the way things get gradually less monolithic. In the past, there was only a single right way to do things—get boiled in oil, thumbscrewed, barbecued or whatever. But gradually, advances in information technology have created more and more things for people to be and do. I'm tempted to risk the wrath of one of our particularly eminent speakers by wondering if there is not a parallel between that and what happens in nature when major environmental change occurs. The most successful organisms survive change because of their ability to develop varieties of themselves. That way, no matter how tough things get, one variety of a species will survive, and through that the species itself.

I wonder if information technology doesn't give us the ability to do that same trick—to become, as a society, more complex, and through diversity, develop a more flexible response to change and therefore survive better. Because if that is true, then what you're going to do over the next 50 years, will cause an explosion of individualism that will put every institution under threat. As one of our speakers will be telling you, institutions are not built for flexibility and fast change. There's a great example from the history of technology of how hard institutions can fight change. All that stuff back in the 12th century when the

European economy recovered from the so-called Dark Ages is generally attributed to the arrival of new textile technology in the form of a new loom. Its main feature was foot pedals that freed the weaver's hands to throw the shuttle back and forth, weave much more cloth, much more quickly and much more cheaply. The well-established European traditional weavers unions smashed every one they could find on the grounds that it would, "put people out of work." It was remarkably modern thinking for the 12th century.

However, a generation later, when the dust had settled, market forces put the loom back in use. The thread makers couldn't keep up and they caused trouble until the answer came in in the form of a spinning wheel from China, which made thread fast enough to keep pace with the wheel and the loom together. The production of cloth went up like a rocket.

But more riots broke out because the mass-marketed cloth was linen, made from plants because they were cheaper than wool. So the rioters that time were from the well-established wool industry. However, market forces meant that everyone eventually wore this new cheap cloth, and when they wore it out, they threw it away. So all over 14th century Europe, there was this gigantic and growing pile of linen rag. That caused the price of paper to drop like a stone because linen rag was the best raw material you could have and it was now free.

More riots broke out in the wool industry again because parchment is sheepskin and now it was too expensive to use. But because there was now enough paper to stick on the walls, it had become a seller's market. The scribes were overworked and in demand, and pretty soon the old established professional writers guild was going on strike for higher wages. The Black Death had knocked off two-thirds of the population of Europe, the other one-third was inheriting like crazy and there was not enough writing ability to go around for all the necessary documentation—that is, until about 1450 when Gutenberg solved the problem by automating it with a printing press.

This caused riots in that greatest of old established institutions—the Vatican. The Pope needed a printing press like a hole in the head because it was thought to encourage free thinking. That changed when somebody realized that the printing press could be used to print indulgences. For those non-Catholics among you, an indulgence was a kind of spiritual credit note: Pay now, sin later. With all the demand for printed salvation that followed, Rome made a million. The money was used to remodel the Vatican, pay Michelangelo's bill and allowed Rome to get involved in some prestige projects that made certain German clerics madder than hell at this cash-and-carry view of salvation. One of those clerics nailed up a few mild remarks on the subject, and there—thanks to advances in textile technology and fought by institutions all the way—is the Reformation.

It's a trifle oversimplified, but you get my drift. Institutionalized thinking doesn't take kindly to unexpectedly new ways of doing things. It's like the lady in the hotel elevator. A man she doesn't know gets in, the doors close and as they start to rise, he says, "Your room or mine, baby?" And she says, "If you're going to argue about it, forget it." What I'm saying is that there is a built-in accident waiting to happen between this industry and what it's going to do to the major social structures over the next 50 years. When a manager in Boise, Idaho, is using a Japanese corporate satellite to run CAD/CAM units in Argentina the way the London accountants say the Taiwan headquarters wants using software uplink live from Sydney, what happens to national sovereignty?

The final area of change to be aware of besides the domino effect of innovation, the key importance of information technology and the way things get more complicated, is what might be called the "user effect"—the way the marketplace can influence the direction of innovation. You don't need to be a Ph.D. to be a user with influence. James Watts's pump might never had done more than drain a few mines had it not been for the surge in the population of those with cash in their pockets, and their desire for metal pots and pans, that kicked off the Industrial Revolution; or the thousands of people who just didn't like the Ford Edsel. These world-changing consumer decisions don't result from debates and focus groups. Individuals make individual choices all by themselves. But if it happens a million times over, it changes everything. Theoretically, that's what we all do at elections.

We have some great speakers for you, and they have some very provocative and exciting things to say about the industry and what it's capable of doing over the next 50 years, and I thank them for taking on this task. Kick off with Gordon Bell, talking about what we all want to know about: The law of prediction. He's followed by Carver Mead, who's going to be looking at semiconductors and whether or not we can expect Moore's law to go on operating through 2047. Next comes Joel Birnbaum with a challenging look at the alternative types of computing that lie ahead and which are likely to predominate. He's followed by Pattie Maes to discuss what might be described as one's future significant other, a k a the electronic agent. Next comes Nathan Myhrvold on the matter of software and who's going to be doing it, or rather, what's going to be doing it. Today ends with fun and frolic—and who knows what—from the world of 21st century entertainment in the form of Bran Ferren from Disney. Tomorrow, we open with a look ahead to whether or not we'll still be there in 2047, with former Secretary of Defense William Perry, who'll be talking about computers and war. Then comes former Finance Minister of Chile Fernando Flores, looking at the impact of information technology on your business and how deals will be done 50 years from now. Next will be Vint Cerf with an intriguing look at a future where the Internet and computers are invisible, pervasive and everywhere. Brenda Laurel follows with a very provocative look at computers in culture that I think will surprise and excite you. Immediately after lunch tomorrow, we have my countryman Maurice Wilkes who wrote the first book on computer programming. Then comes Elliot Soloway on what the computer industry is going to do to education between now and 2047, and a very different world of qualifications that may require. Finally, Reed Hundt, FCC chairman, will talk about what telecommunications is going to do to education and most other things. On Wednesday, we start with a presentation by that well-known science fiction writer Bruce Sterling who, by the way, suggested that the best way to use a polling paddle is to see if you can do the Mexican wave, so you can guess that his prediction is going to be far from predictable. After him comes Raj Reddy, out on a limb with teleportation, time travel and immortality. And last, but very far from least, Noble Prize winner in quantum chromodynamics Murray Gell-Mann will address the complex matter of a convergence of physics, biology and information technology. That's a subject that in itself kind of wraps it all up.

The Folly of Prediction

1

James Burke

Our first speaker is known throughout the industry as the Father of the Minicomputer, although I'm sure he has mixed feelings about this. And I guess he is the embodiment of the saying, "You only know where you're going if you know where you've been." He's been a very significant part of the industry since the very early days, so when he talks about the matter of predicting, his views are grounded in considerable experience of doing just that in his 23 years, for instance, as vice president of R & D at Digital Equipment Corporation. He was educated at MIT. From 1966 to 1972, he was professor of Computer Science and Engineering at Carnegie Mellon, and from 1986 to 1987 was the first assistant director of the National Science Foundation Computing Directorate. He led the National Research Network Panel that became the NIIGII and authored the first high-performance computer-and-communications initiative. Today, he's an industry consultant-at-large and senior researcher at Microsoft, concerned principally with telepresence. So you could say he's been around the block. He's written numerous books and papers on computer structure and startup companies, and in 1991 he published *High Tech Ventures The Guide to Entrepreneurial Success,* so you may want to take notes. Currently, he's on the technical advisory boards of Ambert Adaptive Solutions, Cyrus Logic, DES, Fake Space, University Video Communications and others. He's also director of the Bell Mason Group, supplying expert systems for venture development to startups, investors, governments and entrepreneurial initiatives. His many awards include the Institute of Electrical and Electronics Engineers Von Neumann Award, and the National Medal of Technology for his "continuing intellectual and industrial achievements in the field of computer design, and for his leading role in establishing computers that serve as a significant tool for engineering, science and industry." I suppose what that all adds up to is you're about to hear it from the horse's mouth on a subject we are all desperate to get right: predicting the future. The title of his talk—and this is where you can tell he has a handle on the matter—is "The Folly of Prediction." Ladies and gentlemen, please welcome Gordon Bell.

Gordon Bell

Why Do We Predict?

What is this urge we have to predict? Where does it come from? The desire to foretell has always been with us and is perhaps a selective advantage for our species, since our lives and fortunes have frequently depended on guessing correctly. With respect to social Darwinism, we might say that the ability to predict accurately affords a society a selective advantage. In ancient times, the astronomer priests looked to the heavens to predict the time for the spring planting. Joseph's Biblical prediction of seven lean years after seven fat years led to seven years of saving for that proverbial rainy day. Prediction is also a source of individual power. One who could predict a solar eclipse in an age when such phenomena were feared was an important person indeed. The one who could predict the course of an illness, the next rainfall or the return of the caribou acquired power and prestige.

Here we are concerned with the history and science of prediction as it applies to computing. While such prediction is not usually a matter of life and death, the motivations are the same: profit, prestige, power, safety, privacy and security. At times we engage in prediction merely for amusement or out of curiosity. It enriches our lives by providing a vision of what we can become as individuals and as a society, and by offering us the challenge to achieve our vision. Moore's law, which describes the exponential rate of increase of chip technology, is one such challenge and vision. Another was presented to the scientific community in an article by Vannevar Bush in the July 1945 *Atlantic*. There, he outlined a vision—prescient conceptually if not in the specific technologies—of how knowledge would be organized in the future.

On a more commercial level, it is necessary for corporations to plan—at least that's what business school teaches—in order to avoid surprises. No manager wants to be taken unawares. So they start planning, and the next thing they know, they're given a budget to realize their plans, and now they have to maintain this budget, which requires further planning . . .

Market predictions have become an integral part of the corporate culture. In order to raise capital, there has to be a business plan complete with all sorts of predictions. I find it amusing that this has given rise to a professional class of predictors—a sort of latter-day coterie of sibyls or Delphic oracles. People actually pay other people to predict a new market they're entering, be it pen-based computing or video on demand. They have no idea whether there will be a market for their product, but the oracle has spoken because they paid it to speak that way, and on the basis of that they go out and raise venture capital to make the vision come true.

And sometimes it does. Our predictions can help us extract large sums from the federal government or other funding sources. For example, now that we've got a teraflop, we could predict that if we had a petaflop, we could build a better bomb by next year, or journey to Jupiter in 2001 or perhaps create a virtual Jupiter right here on Earth. If we are creating grand challenges for ourselves, perhaps it is merely to perpetuate our own funding so we can keep on doing exactly what we're doing.

Finally, let's not forget predictions for the sake of idle curiosity. It's fun to make predictions, to guess at what the future will bring. Perhaps it's a way of transcending time, of grabbing a piece of tomorrow—or maybe it's just because I like to make bets. I've never been to Las Vegas, but I get my gambler's frisson by betting on the future of the digital universe. I have a little fun and perhaps earn a little money on the side. In 1996, I bet my friend Jim Gray that there would be videophones in wide use by 2001, and I expect to lose! The bet was designed to encourage me to work to make this come true, or in the words of Alan Kay: "The best way to predict the future is to invent it."

Learning From Our Past Predictions

Since the birth of the computer, many predictions have been made about the future of digital technology, some of which have become well-known. Probably the most famous post-Babbage, post-Turing and pre-computer-industry prediction about computing was made in 1943 by Thomas Watson, chairman of IBM, who said, "I think there is a world market for maybe five computers." Fifty years later we can all laugh at how wrong Watson was, but if we look at the first large-scale calculator that IBM built for Harvard—a 50-foot-long behemoth—we might learn that predictions require some history. Watson had no history of computers on which to base his prediction. In fact, we didn't even call them computers then—that's what we called the folks who did the computing! But his was considered a great prediction because it held for about 10 years. All predictions are at least implicitly time-limited—if nothing else, the principle of entropy will sooner or later put an end to everything. In general, the less historical precedent we have to go on, the shorter the time period of a prediction's validity. With no history at all, we might as well be reading tea leaves or entrails. We need history in order to predict!

Another set of predictions from the early days comes from a report I read for amusement from time to time: the 1969 report of the Naval Supply Command, which employed a panel of experts dubbed "Delphi" to forecast the state of computing over the next 15 years. It is interesting to look back and see what people were envisioning almost 30 years ago. Most of their predictions were wide of the mark, though a few were on target. One of the accurate ones had the use of punch-card readers declining after reaching a peak of 1500 cards per minute in 1974. Frequently, the panel's predictions were hurt by their ignorance of the market—some of the products they were predicting were already out there. So, while their punch-card prediction was correct, they were apparently unaware that in 1969 there were already readers available that exceeded their 1500-card-per-minute prediction. The Delphi panel also predicted that advances in computer memory would give us large memories on the order of four megabytes. At the time, memory was based on magnetic cores. Based on the state of technology, it seemed likely that magnetic cores would still be produced at the end of the next decade. But in the 1970s, metal-oxide semiconductor (MOS) memories appeared from nowhere and, like the comet that did in the dinosaurs, wiped out magnetic core.

In 1972, I gave a future-of-computing talk at MIT. My first prediction was that future computers would be both cheaper and faster, with every decade seeing a new platform based on reduced cost. I also predicted that the semiconductor companies eventually would become the computer manufacturers. This was right after the introduction of the first microprocessor, and I foresaw a time when there would be an entire computer on a chip, including the processor, memory and I/O. This is finally happening, and we now have system-on-a-chip computer companies. After that, I declared that we were badly in need of networks, because we were at the point where people were becoming the networks that transported information between the various computers. If I had carried the thought further, I would have concluded that all computers needed to be linked together, just as the Internet does. My final prediction—one that was considerably less prescient—concerned semiconductor evolution. As I saw it in 1972, semiconductors were going to evolve for six more years and then level off. In fact, the number of transistors per chip and performance are still doubling every 18 months and are projected to continue doing so for at least another decade. This misprediction is understandable because as an engineer, I couldn't see beyond the projects that we were creating for the next two, three-year product generations.

A number of our predictions were overly optimistic. I recall my own 1960 forecast that, without a doubt, speech recognition would hit the scene by 1980. A 1962 prediction by the head of RCA Labs on the commercialization of speech was even more fanciful: He described a speech typewriter that would take in data from a microphone and produce printed pages; then another machine would translate that into the foreign language of one's choice; and finally, a third machine would spew forth the translated version as speech. To put it bluntly, speech predictions have been wildly optimistic and consistently wrong. My prediction that speech recognition would occur in 1980 was made after I had spent a year working on computerized speech. At the time, I said to myself, "I don't want to be in this business; this is a 20-year problem." Well, I was wrong. It was a 40-year problem, which is about where I usually am on my predictions—optimistic by a factor of two.

Another Delphi panel prediction had parallel processing being online by 1975. In fact, the Cray 1 confirmed it. I recall a wager I made with Danny Hillis who was sure that in 1995, massively parallel computers would do most of the scientific processing. His prediction was based on a DARPA (Defense Advanced Research Projects Agency) initiative to produce a massively parallel machine, which might actually have come to pass had the company not gone out of business. A 1,000-node multiprocessor—which did not make it by 1989 as I had predicted or planned and which I was asked to build by 1992—was finally available in 1996. The lesson: Parallelism is more difficult to implement than anyone imagines.

New Overtakes Old, Forever Thwarting Predictors

While an established technology gives us more of the same over time, eventually a new technology, which at the outset may be more expensive and less productive, replaces the old one as its superior qualities become apparent. As the new technology is developed, it becomes more efficient, faster and cheaper through better production techniques and a declining cost curve. While constantly improved performance is taking place at the top, from the bottom come new, lower-priced computers. This model, which I came up with in 1972, still holds true today. Newer and cheaper wins; the old dies off. The mainframe would be dead if the market based its decisions solely on the cost of operations, yet it is still hanging on just to hold our legacy data and run legacy programs. Based on the high cost of converting programs and data, one can almost predict that it will be with us for the next century.

So what can we really predict about computers? A few years ago I predicted the network computer, and my friends at Microsoft and Intel assured me that no one wanted anything that cost less than $2,000. In 1998, the challenge is to produce home computers costing less than $1,000 and network computers for only a few hundred. Then there's the system on a chip around which an industry is now forming that includes processing, memory and interfaces to the real world. The home area network has arrived, and now we're talking about various forms of body area networks—computers that are actually attached to us. Having had two heart attacks, I'm particularly interested in one such item—a cardioplastic implant, whereby a piece of back muscle is wrapped around the heart and becomes part of it. Some experts recently predicted cochlea implants, which are already here. Bionic limbs, whatever those are, will arrive by 2013. Artificial vision will take much longer because all the complex fiber optics will have to be coupled in.

Law I: Expect the Expected

History can be a reliable guide to prediction. Knowing what has happened in the past guides us in predicting how things will progress in the future. There are some cases where the past is completely reliable. Even before Newton propounded his law of universal gravitation, predictions of a sunrise every day and the return of spring every year were made on the basis of their regular occurrence. With computing, as with any human enterprise, predictions are more subtle, but knowledge of past performance can help. Thomas Watson's prediction that there would be a global market for only five computers needs to be greatly modified on the basis of the history over the past 50 years. With historical knowledge, we can create a mathematical model and extrapolate into the future. Perhaps the most well-known extrapolation is Moore's law, which asserts that circuit densities of semiconductors will continue to double approximately every 18 months or, equivalently, increase at a rate of about 60 percent a year. We started around 1972 with 1K of memory, so if we run that out to 2010, that gets us up to 40 gigabytes for a single memory. Moore's law comes straight from the product-development cycle, whereby new semiconductor processes, materials and products yield a new generation of chips every three years. Indeed, the product gestation time for a minicomputer was three years. For a microprocessor it was also three years. A mainframe's was four years, while supercomputers continue to take at least four.

While we've got Moore's law working for us, in January 1997 a single electron was stored in a seven-by-seven-nanometer cell at the University of Minnesota. Based on this achievement, Nathan Myhrvold, chief technology officer at Microsoft, has noted that by 2010 we could get two and a half petabytes on a chip, which would accelerate Moore's law by 30 years. So is Moore's law too conservative, and can it be accelerated?

Law II: Expect the Unexpected

If the future continues along the lines laid out by the past, then we should be confident about our predictions. However, as we've seen over and over, just when we thought we knew what we were about, a new technology comes along to upset historical precedent. So another law of prediction is to expect the unexpected: Sooner or later some cataclysmic event will upset the virtual apple cart. We saw this happen with magnetic-core memory. Core seemed destined to stay for a while until MOS wiped it out. The same thing happened when CMOS wiped out the bipolar logic that was used in minicomputers, mainframes and supercomputers. Cray Research hung on to bipolar too long, and the Japanese built a range of CMOS supercomputers about six years ahead of them, offering faster, lower-cost products. (Cray solved the problem by getting the U.S. government to levy huge dumping tariffs against the Japanese.) The lesson is that over the long term, unexpected events will cause discontinuity and rapid change. Thus, in making predictions we must try to build the unexpected into our calculations.

Law III: For the Short Term, Bet Against the Optimist

We take Moore's law seriously because it has proved to be a good indicator over the last 30 years. Unlike natural laws, Moore's law is about the collective behavior of all those involved with the semiconductor industry—researchers, materials and process suppliers, chip manufacturers and users. A prediction gains credibility if it has historical parallels or has been made by someone with a successful track record. For short-term predictions, bet against the optimists. Their vision is clouded by desire, and what they see as just around the corner might still be a long way off. The further such visionaries are from the reality of actually having to produce what they are predicting, the more overly optimistic they are likely to be. Conversely, if the person doing the predicting is a scientist or an engineer, someone actually at work on the project in question and aware of all the hurdles, they are likely to be quite conservative. Six years is the horizon of the engineer who is on a three-year product cycle. Products under development typically take three years or so to reach the market, so one hears short-term predictions that, in some area, progress will continue for six more years and then flatten out—that is, after two generational cycles.

With respect to longer-term predictions, the development cycles of science and engineering are good indicators. Here we can cite Carver Mead's rule, which states that it takes 11 years from the first observation of some phenomenon to the point where it can be commercially successful. Some examples of this are the invention of the transistor in 1946 to its adoption in a computer in 1957; integrated circuits in 1956 to their adoption in 1967. The first Arcnet LANs were developed in 1972, and adoption of the Ethernet happened in 1981—only nine years later. The time period from invention to adoption is shrinking. From the inception of the Web and the publishing of documents using HTML browsers to adoption in PCs and workstations was only a couple of years. With the Web, software now spreads like wildfire.

Law IV: Don't Assess the Market Based on Your Own Personal Characteristics

We can learn about the laws of prediction by examining failed predictions. Furthermore, it's gratifying to see those people we thought should have known better landing so wide of the mark. A former boss of mine, Ken Olsen, the president of Digital Equipment Corporation, predicted in 1977 that there would be no market for home computers. At that time I had been studying home use of computers among colleagues at DEC for about 10 years. He more recently predicted that his company, Modular Computer Systems, would not be on the Internet, which was not a very good prediction, since at the time they already had a home page. Out of such predictions we may infer the following law: It is a mistake to equate yourself with the average user unless you happen to be the average user, which no one is. On the other hand, computing has progressed very rapidly because the designers have been developing systems that they want to use in their work—essentially becoming producer and consumer simultaneously. However, since these designers are decidedly not the average user, a frequent result has been the implementation of user-unfriendly interfaces of the type Unix users know all too well.

Law V: Predict with Exponential Data when You Have a Few Data Points

When we have historical data on some event, we can fit it on a curve and extrapolate it out into the future. Of course, the data can be put on more than one curve, and the less data, the more freedom we have in deciding what curve fits best. This is particularly true with exponential growth. Exponential-growth curves appear almost flat for a long while when plotted on a linear scale, then the growth seems to become linear and suddenly takes off. Therefore, we have to employ the exponential plot with care, using it only when the change is truly exponential, which happens with observed growth history, underlying production capacity (learning curves) or consumption (demand curves). On the other hand, if we're too conservative, we'll miss the boat. Sometimes, what looks like no growth or linear growth may merely be the slow-growth part of an exponential curve that will eventually take off. The growth of Internet traffic is a good example of that. We didn't see the huge growth in the Internet, prompting Bob Lucky, vice president of Bellcore, to remark in 1995, "If we couldn't predict the Web, what good are we?" The growth of Internet use was exponential all the time, but the slow part of the curve looked linear, particularly since nobody anticipated the invention of the World Wide Web—addressing, HTML and the browser. The Web appeared just in time to maintain the Internet's exponential growth.

Internet growth has always been exponential—doubling every year. That allows us to project its growth based on its history. We can predict, then, that there will be a crossover point somewhere around 2003 when there will be more Internet users than there are people in the world. Obviously that won't happen, but it needn't necessarily dampen our prediction, because in figuring Internet growth we may, by then, have to factor in the auto market, the light-switch market, the camera market and who knows, maybe even the dog and cat collar market. The year 2003 prediction was based on four data points: the number of current Internet users, the observation that this number is doubling every year, and the current world population and its rate of growth. Nicholas Negroponte, head of MIT's Media Lab, predicts that by 2000 there will be one billion people using the Web. I think his estimate is a little bit high. That's more people than there are PCs and it's about the number of TVs and phones, so the current infrastructure will not suffice. We'll need something like radio links to get there. So I don't think its a good prediction, especially since I've bet him $1,000 even money that by December 31, 2000 there won't yet be a billion Web users—each with their own address—and five-to-one odds that there won't be a billion by the end of 2001.

I prefer to use just a couple of data points for predicting exponentials, otherwise unnecessary and usually unhelpful complications are introduced. So with something that you believe is growing exponentially on the basis of technological development, the economy and the marketplace, two data points will suffice. More points simply don't add more information—it's like computing more digits in a measurement that exceeds the accuracy of your measuring tools. For example, consider the growth in processing speed. We can begin with Maurice Wilkes's 1949 EDSAC computer which processed 700 instructions per second. In 1951, there was a jump to 50,000 ips when parallel memories using CRT storage tubes, and eventually core memories, replaced serial delay lines and drum memories. After that, growth continued exponentially, and depending on what we use as our second point, we can predict from 20 percent to 40 percent growth per year. Growth in memory and storage has been similar, and in fact, network capacity parallels all of these. The slowest growth component of the network is telephony at about 15 percent per year. It seems, therefore, that the limiting factor is going to be our ability to access the network from our homes.

Law VI: The More You Spend, the More You (Might) Get

An important factor in the growth of a technology is the amount of money spent on a computer because, in general, the more you spend, the more you would like to get. So, if someone is willing to spend the money, you might get better results sooner than later. In the 1960s, computer pioneer Herb Grosch declared that computing power grows as the square of what is spent for a single computer. This law has been refuted many times, with most observers believing instead that performance grows as the square root of what is spent due to the extraordinary growth in the power of inexpensive microprocessors. Triggered by DARPA's Strategic Computing Initiative in the early 1980s, there have been wonderful results in high-performance computing. A 2,000-fold speedup in the most expensive supercomputers occurred between 1987 and 1997. We can retrospectively predict this by Moore's law (a factor of 100); by spending more (a factor of two), since instead of spending a mere $30 million, we now spend $100 million dollars for the largest supercomputers; and finally, by the switch from custom-designed, low-volume ECL circuitry to high-volume, high-performance CMOS (Complementary Metal Oxide Semiconductor) microprocessors (a factor of 10).

In the mid-1990s, the Department of Energy's Accelerated Strategic Computing Initiative set a target of 2010 for petaflops—1,015 floating-point instructions per second, which will require another 1,000-fold speedup. As we've seen, we expect to get a factor of a 100 with Moore's law. A factor of two comes from spending more for one system. Now, instead of capping spending at a mere $100 million, we can spend $500 million for a massively parallel, scalable supercomputer. If we centralize all three DOE centers into one location, that boosts it by another factor of three; or alternatively, the network can be made fast enough to achieve this speedup. Increasing competition would perhaps generate another factor of three, but this has to occur based on the use of high-volume components. So petaflops by 2010 seems possible in theory. But in a funny way, our quest for the ultimate parallel computer is like parallel lines—they never meet. Our reach eternally exceeds our grasp, and the goal is always a constant distance away. Bill Wulf, president of the National Academy of Engineering and professor of Computer Science at the University of Virginia, predicts—or rather has a vision—that "one can imagine millions of hosts in a loose confederation that to users will look like a massively parallel desktop computer." This is exactly what the Internet is. Harnessing it for a single problem is the trick!

Cyberization Is Our Quest and Our Fate

What can we predict about the future of computing based on our understanding of the history of computers? We have to consider three main factors: the platforms, the networks that connect them and cyberization, which is the entire process of putting information into a computer. It includes all the components of a user interface such as WIMP (windows, icons, mouse, pull-down menu). It includes implants that couple a computer to a person—sensors, effectors, cameras, digital money that encodes bits. It includes a digital representation of the physical world—books, money, newspapers, stocks, pictures and eventually television. Fundamentally, cyberization is the process of making the whole world digital instead of analog so it can be available anywhere, any time—the essence of cyberspace. Finally, it includes "state" information about all the networks—highway networks (traffic at any point), power grids and water. This is where we're heading. I make this prediction based on the belief that cyberization has become our human and scientific quest to understand and achieve. We see that it is possible, and so it becomes our goal and fate. So cyberization is the interlinking of all our experiences and artifacts into a fractal network of networks of networks that start with the universe and end at our cars, our homes and our bodies.

Networks: Structure and Content

What will be the nature of this network? Will it be one network for data, a second network for such things as telephony and then other networks? Or will they merge into a single dial tone? I think we'd all like to have the latter.

Bandwidth Is Free, but None Is Available at That Price

There are many factors that influence the evolution of the network. Chief among them are commitment, money and vision. Irwin Dorros, who was head of AT&T Long Lines in 1981 before the breakup, thought that integrated services digital network would be ubiquitous by 1985. He was wrong because at the time there was no need for such a network. If ISDN ever becomes the ubiquitous connection, it will clearly be by default. It's not something any of us are anxious to have, but in fact, it's going to be in the communications line. ISDN was an expensive investment that had no application and was never finished. When it should have been deployed for computing, the telephone companies didn't do it. Now that we have the computers to connect, the bandwidth is too low for video (by a factor of 10 to 20), and the presence of ISDN has inhibited investment in a faster network. Furthermore, conventional POTS (Plain Old Telephone Service) is running at half the speed of ISDN, so evolution is catching up with what was to be revolutionary simply because of economics and commitment of deployment. I have worked in this area a long time, and from my experience I can say that network bandwidth becomes available more slowly than anyone can ever predict. So, beware of predictions based on the notion that bandwidth is free. It's free alright—you can't get any of it at that price.

Vision and Faith in Science

Now for a prediction I'm happy to make. It goes back 50 years to Vannevar Bush's *Atlantic* article in which he wrote, "There will always be plenty of things to compute in the detailed affairs of millions of people doing complicated things." So, I feel I can safely predict that we won't run out of things to compute, which means that those of us in computing have chosen a wonderful area. On the basis of Bush's predictions, I conclude that faith in science and a vision of what can be useful are good predictors. In his article, Bush outlined an information storage and retrieval system with hyperlinks that he dubbed "memex," which many believe was the original vision for the World Wide Web.

Of course, in making predictions it helps to be a Vannevar Bush. It also helps to be really lucky, because none of the technology that Bush outlined turned out the way he saw it. Bush had been responsible for technical manpower during World War II, so he had seen some amazing developments including radar, the jet engine and various automatic control systems. But all of these inventions pale in comparison to the integrated circuit and the computer. With the amazing things that have already been done with them in such a short time, and the even more amazing things—many as yet unimaginable—that will be done with them in the future, they will, I believe, turn out to be among mankind's greatest inventions.

CARVER MEAD

Life Without Bits

2

James Burke

Our next speaker is one of the industry's greatest teachers, so when he finishes there will be a test. He's the Moore Professor of Engineering and Applied Science at Caltech, where he's been since he earned his Ph.D. there in 1960. Now he's a member of so many associations, I'll just mention a few: The Institute of Electrical and Electronics Engineers (IEEE), the National Academy of Engineering, the National Academy of Sciences, the Royal Swedish Academy of Engineering Sciences, the American Academy of Arts and Sciences and the Franklin Institute. His many awards include the IEEE Centennial Medal for extraordinary achievement and its Von Neumann Medal for "leadership and innovative contributions to VLSI (Very Large Scale Integration) and creative microelectronic structures." In fact, he pioneered the field. Today at Caltech, he focuses on modeling neuronal structures, and his latest book is entitled, *Analog VLSI and Neural Systems,* so I suppose you could say he does a great deal of thinking about thinking. That's an English joke—I won't do another one. I have no doubt, however, that thinking is what you will be doing after you hear what he has to say, because he's going to speak about the future of semiconductors. That's a subject so close to everybody's forecast charts—how long will Moore's law persist?—a matter that may have some bearing on what you or your company may or may not be up to in the relatively near future. Please join me in welcoming Carver Mead.

Carver Mead

Computation started with a belief—basically, a conjecture—that any computation could be done by this funny thing we call a Turing machine. It's really just a model of computation, something made up in a person's head. It was actually a model of mathematicians proving theorems, so that's an interesting thing. I have a slightly different view of computable functions. I think computable functions are really the things you can sit down and run on your PC or workstation or whatever. I'm not the first person to think this. That's an old idea and there's a whole science of complexity theory which deals with all that.

What do we believe about computing? Well, at least from a scientific point of view, we use the word "time" to mean the number of steps it takes on our computer to do something. By space, we don't mean space in the physical sense; we mean the amount of memory it takes to do something. The underlying assumption is that all computers are basically alike within some small factor. And once we've defined computers to be all alike, except for the year in which they were made, we can define a set of problems that are tractable—namely, the ones we can run on our PC that get done before dinner. The intractable ones are the ones we run over the weekend and they still seem to be running—but we're not sure. Those problems are characterized by a very large number of alternatives, exponential in the worse case, and the lack of really good ways to make shortcuts.

This conference is about congratulating ourselves on how we're changing the world and all that, but I would like to call attention to the fact that there is an interesting class of problems on which we're not doing so well. Most interesting optimization problems are like that, but what's really interesting is that *all* the perception problems are like that—hearing, vision and the like. That wouldn't be so bad if none of us could do them on a computer. You heard in Gordon's talk the failed predictions about any of the perception problems. They are still failing as we speak. It wouldn't be so embarrassing if it weren't for the fact that even very dumb animals like the fly do them very, very well.

But, what if we could make a different kind of machine? What if we could make a machine where the computational capability got exponentially bigger with the size of the machine, instead of linearly—or slightly sublinearly—like it does today. That would change the game, because we would no longer be stuck with this device where we continue to make these predictions—speech recognition, artificial vision, artificial intelligence—and they fail because these problems are exponential. My colleagues always caution me that many of these problems aren't really provably exponential—they may just be a very high power.

Well, this talk is a personal confession of mine. For my last 30 years of work, I have been searching for some kind of machine that would do something like this. That wouldn't be just one more machine; that would change the nature of the game. There are some candidate structures; ultraparallel VLSI might do something like that; neural-computing structures we know do something like that, because we use them to watch and listen to each other. And then there's a furor over quantum-computing structures. I worked a long time on ultraparallel VLSI structures and, as in so many things, there's some good news and some bad news: The good news is that we finally learned how to design this stuff to where it sometimes works on first silicon and usually works on second silicon, and they do some things very, very well, and that's all very nice; but the bad news is that the speedup is polynomial at best and usually linear in the amount of circuitry.

So, what's going on? Well, let me once again insult everyone's intelligence and remind you about digital systems. What is a digital system after all, and why is it a good idea? It represents information using a finite set of discrete symbols. That's not a new idea; it came from the invention of the alphabet around 1500 B.C. So why is it such a big deal now? Why did it take so long? It's a good idea for a very simple reason. Let me give the simplest, most brain-dead example: If we have some information that starts out with a one or a zero, it has to get stored somewhere; it has to get transmitted through something and it has to be received by something and so forth. That physical stuff has its own properties, which tend to take nice sharp distributions (the ones and zeros) and turn them into broader ones. If you keep doing that step after step, the distributions get broader and broader and finally overlap. Ultimately, you lose the information. To keep from losing it, we put the signal through a contractive mapping and it comes out the other side as good as new. So that's the big deal about digital. There are more complex examples like error-correcting codes and all. But they rely on the same discreteness of the underlying symbols to do that.

With that art form, we congratulate ourselves on being able to reconstruct the information perfectly—and rightly so. But the form itself carries some limitations, and if one wants to go beyond the limitations, maybe one ought to look at what the limitations are. What we mean by time is the number of steps that some machine takes to perform an operation. But it doesn't really have much to do with time. In fact, these machines don't really have a representation for time, per se, in any natural way. The continuous variables, of course, have to be represented by numbers—strings of these discrete symbols—and they have to get processed in discrete chunks, one at a time. So in the digital world there's no notion of locality or continuity that exists in the physical world, and much of what goes on in the physical world is really simplified a great deal by the continuity that exists there; that continuity is lost when we digitize the information. Therefore, all the alternative hypotheses for the solution to one of these exponential problems have to be spelled out in these discrete symbols without any natural continuity between them, and for that reason we work them one at a time, one after another. This comes back to the fact that the power of the digital system is also its Achilles' heel. The fact that we quantize after each very simple computation means that we lose that very important continuity, which is a representation of physics—the nature of the world out there.

This wouldn't be at all embarrassing, except there are systems that handle this kind of problem very nicely. Ramón y Cajal's original book on the histology of the nervous system illustrates the typical kinds of nerve cells in the brain. These are structures which solve the kind of problems that we have broken our picks on through the last 50 years, without really having made a dent in them. If you look at one of these neurons, it's rather strikingly different from our standard digital computer. It's made out of goo, and that's not quite so striking, but what is striking is that it has on the order of 10,000 inputs instead of the two and a half, plus or minus one or two, that our standard logic gates do. Spines exist on the side where the inputs enter the neuron. Those inputs are clustered rather far from where the quantization takes place down in the body of the cell. How does all this marvelous computation happen? The inputs come in as pulses that are discrete in time and discrete in amplitude. But the relative time of arrival among the nerve pulses coming in is a continuous variable, and that's where the information is encoded. Certain combinations of nerve pulses arrive in the distant parts of the neuron, and they propagate down through these active processes. There's a distributed amplification that goes on there, which has to be done just right. If you turn up the gain a little too much, those little signals coming down grow up and become just another nerve pulse, and they quantize too soon. If the gain isn't enough, the little signals coming down get lost on the way, so the thing has to be tuned just right.

We still don't know how to do that with artificial circuits, but we're working on it, and it looks like it's possible. What happens in a structure like this isn't known with great certainty at this point. All the little signals that enter various places at various times propagate down with just the right velocity, so they come together at these junctions with some sort of nonlinear interaction. This structure actually keeps alive an exponential number of possibilities, all at the same time—and that's exciting. That may mean that the structure's computational capability grows exponentially with its size.

This is why I've spent the last 15 years of my life trying to understand these neurons and trying to build models of things that work like them. After all, the nervous system uses electrical signals. We can make electrical signals, too, with silicon, and they have all the same continuity properties and gain possibilities that exist in the nervous tissue.

We've managed to build some structures using this principle. Gordon mentioned that it would be nice to have an existence proof that you could do something even remotely related to what you set out to do. Well, we've done some things that are remotely related to what we set out to do. It turned out to be much harder than we thought to construct anything that even remotely looks like the nervous system. We've built some retinas, we've built some things that look for motion and stereo and that sort of thing. We've built some cochleas, and some things that might be really good cochlea implants. We've built some systems that allow the whole thing to learn as it goes along. As I mentioned, the thing had to tune itself up, otherwise it would become wildly unstable resulting in massive epilepsy, or all of its signals would die out.

Now it's possible for learning to be distributed throughout the system, which works in continuous time. One can imagine a system that is one level of our nervous system. Conceptually, it might look something like this: An input comes into the system in real time. The system contains some kind of network of artificial neurons whose job is to predict what the inputs are going to do next. There are some obvious evolutionary advantages to being able to predict what happens next—not just for companies, but for individuals as well. Then you can look at both input and output and ask, "Did I make the right comparison? Did I make the right prediction?" It's just like Gordon did, except you're doing this on a millisecond-by-millisecond basis. If you've made the wrong prediction, you do two things: You correct the model a little bit, and you send the output to the next level of the nervous system. It's well-known that the nervous system works by only bringing things to your attention that are interesting—not just routine. So the nervous system is interested in things it's not predicting rather than in the things that are all old hat. It's a little like attendees at conferences. When this kind of system comes into being—probably in another 10 years—we'll be able to build systems that look at the world in real time, extract the interesting information and filter it through several levels until the interesting stuff comes to our attention. That's one possibility for a computing paradigm that's very, very different from anything we know today, and which has the possibility of directly addressing some of the hard problems that we really haven't touched.

The other possibility is to use a quantum system, and there's a lot of noise being generated about this today. So what is quantum computation? It involves taking some quantum system that preserves the phase of the wave function of the electrons. You can do this with a bunch of atoms, you can do it in a superconductor or you can do it in a lot of different ways. (No one way works in any real computational sense yet, but it's conceivable that one could be developed with the right ideas.) Then, you encode the information in the wave function of the electrons. If you did that for one electron, it's not very interesting; it just goes around and does the same thing. But if you encode it in the wave function for many electrons, the wave function can be described in some kind of space; and the coupled wave function of many electrons is in a space which is the Cartesian product of the spaces for the original ones. This allows you to enormously enlarge the space within which this collective system is evolving. (That sounds like an exponential to me.) Then you watch this time evolution. Of course, you have to be careful with quantum systems, because if you watch them too much they get nervous and do something stupid.

But if you're careful about all this, in principle you can do computations of this exponential character, the most famous one being encryption. The most well-known computation to date shows that the factoring of large numbers—upon which many of the public key encryption systems rest—is actually a very easy problem if you do it with a system like this. In principle, it can be done in a very short time, even for very long numbers. No one has built a quantum system which actually does this now, and there are lots of inherent problems, but as Gordon noted, if you can see a principle that's based in physical reality, at some point we're going to find a way to embody it in a physical reality.

Meanwhile, computer science theorists now have a whole new model of computation to play with—and they're having a blast. It's providing a much better hands-on, working-level quantum mechanics rather than pie-in-the-sky theory. Still, we don't see anything that is, in principle, a complete showstopper. The practical problem is that the wave functions of electrons tend to get coupled to other stuff in the universe, and when they do that, they tend to lose their coupling with the ones you want. The result is that information can get lost, and right now that's the biggest concern about quantum computation.

In summary, I would say that digital systems look like the whole world to us today. I look at them as the first step toward the development of a range of computing paradigms that will be with us 50 years from now. Then, I suspect, we won't be using "computer" and "digital computer" as the only words that describe computation. We'll have other procedures that we recognize as computation, like the things that are done on the nervous system and the things that are done in quantum systems; we'll recognize them as interesting forms of information processing. Communication on the network won't be the only thing that's interesting; we'll actually be doing more interesting problems with the real world, with real time, with real perception and with real understanding of the world around us. We'll have many more ways to get inputs from the natural world into our cyberspace, and that's all going to make for very exciting 50 years.

Alternative Computing

3

James Burke

Joel Birnbaum is somebody you could describe as a mover and shaker because of what he does, and where he does it. He started out with a doctorate in Nuclear Physics from Yale in 1965, then spent 15 years at IBM's research lab in Yorktown Heights, New York, where he became director of Computer Sciences. In 1980, he joined the company he's been with since, Hewlett-Packard, as the founding director of its computer research center in Palo Alto, California. Along the way he's been vice president and general manager of the Information Technology Group. Today, he's the director of Hewlett-Packard Labs and is HP'S senior vice president of R & D. He's a board member of both the Corporation for National Research Initiatives and the Technion University of Israel, and he serves on the advisory councils of UC Berkeley, Stanford, Carnegie Mellon and Yale. As you can see, a lot of people think he knows what he's talking about. Throughout his career, his personal research contributions have been—and continue to be—in distributed systems, real-time data acquisition, analysis and control. He was one of the early and real pioneers of Reduced Instruction Set Computer (RISC)-processor architecture. You could say that his job today is what this conference is all about—making decisions based on the alternate directions in which the computer industry could or should go. These are decisions on which—in his case and that of his company—a great deal rides, to put it mildly. So not surprisingly, this morning the title of his talk is, "Alternative Computing," which considers the different kinds of computing that lie ahead over the next 50 years, the possible winners and losers, and the kinds of things we're likely to be using them for. Ladies and gentlemen, please join me in welcoming Joel Birnbaum.

Joel Birnbaum

I think it should be obvious to everybody here that you shouldn't really expect technologists to be very good when they pretend to be Jules Verne, Leonardo da Vinci or H.G. Wells. We haven't been very good predictors of the future, and if technologists aren't good at it, what should you expect from a manager of technology? After all, what we do is try to understand the needs of users today, apply some imagination and then see whether or not our current technologies can be extended to meet those needs—or if not, whether we can produce some variation of them. If we're lucky, as you heard very eloquently from Gordon, we can perhaps look ahead 10, maybe 15 years, but certainly 50 years is far too long to use an extrapolation technique. That's why the science fiction writers are so much better at this, because they think of what they'd like the world to be, and not how to move technology incrementally to keep the profits up.

Nevertheless, I enter this talk with a great deal of hope and optimism for the future. It's a trait I've developed by working in big companies, but it's also because I've noticed that as technology marches along, we often see a replacement phenomenon—what I think of as an x-less y. For example, when cars came along, you could think of them for a while as horseless carriages—a carriage without the horse. The same goes for wireless telephones and so forth. So one of the questions that I'll try to address is what will be the x and what will be the y in the interaction between computers and humans 50 years from now. In fact, what I'm going to look at are the alternatives to electronic stored-program computing, as we've practiced it for 50 years, that have disruptive potential—technology with so many advantages that it displaces the technology that came before it. The integrated circuit did this to the vacuum tube, as have many other technologies we've come to take for granted such as the electronic calculator, which displaced the slide rule.

Instead of doing this in my imagination, I've chosen three—quantum computing, DNA-based computing and optical computing—that are already the subject of worldwide research. I'm going to explain how they work, their advantages and some of their shortcomings. Then I'm going to follow the lead of other speakers and throw caution to the wind, shred whatever remains of credibility by that time and predict what we might do with this kind of computing power were it to become available. There's some solace in knowing that when the time capsule is opened 50 years from now, I probably won't have to be there to look at the predictions. I do agree, though, that in order to look forward to the next century, it's a good idea first to look back and see how far we've come.

ENIAC, the starting point for last year's 50th-anniversary meeting, was arguably the first digital stored-program electronic computer. It was also the most powerful computer on Earth for nine years. Let's look at some of its vital statistics: It had 19,000 vacuum tubes, 1,500 relays and weighed 60,000 pounds. It filled 16,200 cubic feet, needed 174 kilowatts of power and could add what was then unthinkable: 5,000 numbers per second. *Popular Mechanics,* in its March 1949 edition, assembled a panel of experts and asked them to project the future of the ENIAC. They, indeed, did some bold things, suggesting that it might have only 1,500 vacuum tubes (about a factor-of-10 improvement), weigh only one and a half tons but still be as powerful as the original ENIAC.

As with most of these kinds of predictions, the error was shocking. Instead of making a machine the size of a sports car with energy consumption to match, we have machines like the latest HP Palmtop computer, which weighs seven ounces, runs for several months on two AA batteries and can execute millions of instructions per second. So one lesson we should learn about the future is not to be too shy about our predictions. We have to understand how and why we made the progress we have, but as others have said before me, we must be sure that anything we propose obeys the laws of physics—conservation of energy, thermodynamics, relativity and so forth.

But there's a second and much more important lesson to keep in mind: The reason these predictions seem so offbase to us today is not that the predictors were stupid. If we were still building machines from vacuum tubes, they might not be too far off. But, in fact, a disruptive technology—the integrated circuit—came into play and changed everything. Furthermore, with the Web and the Internet propagating scientific results almost instantaneously throughout the world, we will see an enormous acceleration in the sharing of knowledge, one that will push progress even faster. Consider, for example, if the Egyptians, while planning the pyramids, had suddenly received a Web message telling them about the wheel. What might we have today?

I'm going to start as others have, keeping the ground rules of conserving physical law in place, and take a look at another depiction of Moore's law as one estimate of what could be in store for computing in 2047. Let's not pay attention to Gordon's notion about the folly of such extrapolations—although I certainly agree with him—and just imagine that we can keep this exponential growing. What would we have? The memory and processor would probably be able to hold about 2×10^{16} bits of data. That's roughly the storage capacity of 100,000 brains. The single data processor would have the processing capability of several million Pentium Pros; memory and processor would have to fit into one cubic centimeter—about the size of a sugar cube—to follow the constant maximum speed of light law. Its power would probably be equal to the combined power of most of the computers that have been built up to today.

This seems fantastic and silly, but it doesn't violate any physical laws, and no matter how improbable it is, it seems possible for clever engineers to figure out how to overcome the barriers to building this computer as they overcame the barriers of optical lithography. What could you do with it if you could? One device of this sort, a nanoprocessor, could do a real-time simulation of the entire Earth's weather with a resolution of about 50 meters. That's pretty good, but not good enough to really test the idea that a butterfly flapping its wings in China might really cause a hurricane in the Gulf of Mexico. Following Moore's law, we would have to wait another 15 years before we could do that calculation.

But before we get carried away with the extrapolation of Moore's law, it's important to say that the engine that brought us to this point, CMOS, can only get us part of the way there. The Semiconductor Industry Association has laid out a roadmap that sets as a goal the continuation of CMOS through 2010. Part of the reason they chose that date is that, by then, all the individual transistors in the circuits will be turned on and off by only eight electrons impinging on the gate—as opposed to the roughly 1,000 that do it in today's technology. Sometime after 2010—certainly before 2020—we're going to run into our first real physical limitation: less than an electron appearing at the gate. Today, electrons and transistors follow the rules of classical mechanics, so what we have gotten used to—the statistical action of electrons—will no longer hold. In spite of this, a lot of researchers are blindly optimistic that this limitation will be overcome somehow.

Maybe. But I'd like to consider instead the first alternative approach to improving the capability of electronic circuits—so-called quantum computing. Now I do this with a fair bit of trepidation, and a short story may make you see why. There once was a gentleman who had survived the great Johnstown flood, and for the remainder of his life, he achieved some local celebrity by telling everybody about the resourcefulness that enabled him to do it. When he finally got to heaven and was asked, as a first-day visitor, who he would like to welcome him, he said, "Well, I really would like it if you assembled all the people so I can tell them about how I survived the Johnstown flood." St. Peter replied, "Fine, but please remember that Noah will be in the audience." Well, I'm going to talk to you about quantum mechanics while one of my heroes, Murray Gell-Mann, may be in the audience. It's a little scary.

The basis of quantum computing lies in the behavior of electrons when we constrain them to dimensions of nanometers at room temperature. In that situation, instead of behaving like classical particles, they behave like waves. We're very comfortable with classical mechanics, because we've interacted with the macroscopic classical world during essentially all of our evolutionary journey to the present. But when we enter the domain of the very, very small, the laws of quantum mechanics rule the behavior of matter. Those laws do not seem to conform to our notions of common sense, which is gained by our experience with the macroscopic classical world. The mathematical constructs that form the basis of much of our understanding of how matter behaves at the atomic and subatomic levels are hard to explain in words. Quantum behavior, though, essentially limits the kinds of things that a device designed for classical operation can achieve. But it also provides tremendous opportunities for the operation of entirely new kinds of devices.

One way to think about this is to characterize electrons as waves. Waves can interfere with themselves where they pass through two openings or travel down two or more wires. This interference could be the basis for new types of devices. Currently, we are making those devices with micron lens scales and we have to cool them in order to see these effects. But when we learn how to make nanodevices that operate at room temperature, the quantum effects will be clearly visible.

How could we use them? Well, one of the things we could do is create a device called an interferometer, which would send waves along two different paths that intersect each other at some region of space. We consider that to be a gate. That means if we launched a wave into the interferometer, it would split up with part of the wave going along one branch, and the other part going along another. When the wave met itself again where the paths intersect, the two pieces would either be in phase—in which case it would grow larger or out of phase—in which case the amplitude would be small. By changing the length of one of the arms of the waveguide, we could change the phase by which the pieces of the wave interfere. By switching the amplitude of the wave, we could build Boolean devices capable of binary logic.

One major obstacle these devices are going to face is the difficulty in wiring them together. One potential solution is to use architectures that need little or no wiring. Examples of these are the many types of quantum cellular automata that have been proposed—charge-coupled devices, shift registers and so forth. They work by passing information—usually in the form of a single electron—from one cell to neighboring cells. Of course, these cannot compete with CMOS devices for high-end computing and storage applications. But if the cells became small enough so that we could fit many billions of them per square centimeter, then their sheer numbers and the speeds at which they could exchange information could make them very attractive. Such structures, made at HP Labs have been called quantum dots. These tiny islands of germanium on a silicon substrate are only 15 nanometers high and have a deviation of only one nanometer. They can trap a single electron in each dot. The major benefit is that these structures are built by chemical self-assembly, a very inexpensive manufacturing procedure and one of the possible ways around Moore's second law, which states that the cost of the fabrication facility is a function of the density of the devices being constructed.

So far, we've only considered performing computations with traditional Boolean logic. Currently, all computers are based on this. But quantum devices provide the theoretical possibility of a new type of logic—so-called quantum logic—that uses quantum bits, or qubits. This logic depends on the paradoxical property of a quantum system known as superposition. Superposition means that sometimes the gate in a quantum-logic system is in the true state and sometimes it's in the false state, but most of the time it's somewhere in-between—a kind of very-small-scale schizophrenia. This state has a nearly infinite number of possibilities. In principle, this peculiar characteristic can be exploited, and it means that we can introduce extraordinary parallelism in certain types of computation. A gate can only be true or false, according to Boolean logic. But in quantum logic, it can have a great many possibilities. Classical computations are done by performing Boolean operations one after another in series. Quantum logic lets us take a huge set of numbers and do all of these operations almost simultaneously. Thus, in principle, a computer that uses quantum logic can use one circuit to compute many numbers simultaneously. In fact, the number of bits in the calculation is represented as 2N, where N is the number of quantum gates in the system. Therefore, if you could build a quantum computer with 800 gates, for example, it would be able to do simultaneous calculation on more numbers than there are protons in the known universe. Even a small quantum computer could take on problems of exponential complexity that we can't even attack today.

Computer scientists have long measured the efficiency of an algorithm. In computing complexity time, we know that so-called nondeterministic polynomial problems are intractable on conventional machines once the number N becomes large enough. If we could find algorithms that would lend themselves to this type of quantum computing, and if we could build quantum computers, then we might see an exponential increase in simultaneous computing and in the type of problems we could attack.

The pioneering work in this area was done by Peter Shore at Princeton a few years ago. Perhaps this example will give you a feeling for how it might operate: Consider factoring a 100-digit decimal number. Even with the kind of computer in 2047 that we postulate would come from the extrapolation of Moore's law—a machine that could do the arithmetic at a rate of 10^{10} divisions per second—we would need 10^{40} seconds to factor a 100-digit decimal number using a trial division method. When you consider that the age of the universe is only 10^{17} seconds, this seems a long time to wait—even for patient industrial managers. That's why encryption codes use large numbers, and that's why they're considered so secure. Now, though, a quantum computer could, in theory, factor such a number in about one minute, although this would require overcoming enormous obstacles. These obstacles, though, don't appear to be fundamental; they appear to require extraordinarily difficult engineering solutions.

All of this is very nice in principle, but how could we make it work? How do we get the information without disturbing the state of the systems? How do we actually build such a device? How could the power for such incredibly dense devices be reduced to practical levels? Do we need reversible logic, for example? How can programming be done? How can we deal with the unavoidable errors in systems that have such vast numbers of devices? Can we service such a machine? There are many more daunting problems. I doubt very much that any one group or any one idea is going to solve all of them. But with the flow of information in the world today, and with so many clever people working on so many questions, over the years I think we'll find some solutions from a combination of current ideas or from a brand new one.

Another problem is finding the algorithms that will let us take advantage of this exponential property; yet another is determining what kind of problems they could attack. Obviously, this could be used to simulate quantum-mechanical systems. It could do Fourier transforms, which would help tremendously in the area of pattern recognition and might enable perception-based computation. Quantum computing has, in principle, an inherent property that makes it ideal for search-and-use databases. Let's say you wanted to search the entire Library of Congress to find an obscure quotation from a single book. Classical computers would have to take that quote and compare it serially with all the quotes in the library. A quantum computer, through the superposition and spectral-pattern matching, would be able to find the quote in a single pass, although retrieving it would not be simple. There are about 30 groups in the world working on building fast, Boolean-logic conventional computers using these kinds of devices. There are at least seven—maybe 10—theories on how to build such devices. A single-electron transistor is one of those, but the vast majority of work here is theoretical. There are several experimental demonstrations of the simplest unit of the quantum computer, which proves that we can do it, but the enormous complexities of trying to deal with this are still before us.

Now I'd like to examine the second alternative: DNA-based computing. In November 1994, Dr. Leonard Adleman of the University of Southern California startled the computer science world with a paper dealing with how he had used 100 microliters of DNA molecules—that's about one-fiftieth of a teaspoon—to solve a simpler variation of the classic "traveling salesman" problem. Since then, many other researchers have taken up the cause, and now a number of them believe that DNA-based computing might solve some problems that are intractable on today's most powerful supercomputers. By virtue of massive parallelism and exponential speedup, such machines are interesting for the same reason that quantum computers are interesting. Remember that while conventional computers represent data in binary digits, these first DNA-based computers use strings of DNA base-pairs. They perform their calculations by carrying out common, biochemical manipulations—combining, copying and extracting the strands of DNA. In other words, you synthesize particular sequences of DNA using ordinary techniques, let them react in a test tube, then you examine the results.

The problem that Dr. Adleman approached was the one he considered in this example: He took seven nodes, each representing a city, and posed the classic problem of how to start in one node and end in another along a path that enters each node only once. He solved it by creating short DNA pairs to represent each city and the route between them, and by first designing the molecules so they would randomly join and test all the possible solutions.

Let's look at it in more detail: In step one, he generated random paths by randomly connecting the pieces of DNA he had previously constructed in unique sequences to represent the seven equal-length chains to each city. This results in chains of different lengths that represent all the possible paths among the cities. In step two, he used a biochemical manipulation—so-called polymerase chain reaction, or PCR, to select only those chains that start at node zero and end at vertex six. In step three, he used a different procedure known in the trade as "PAGE," which is a type of gel electrophoresis operation—it selects only those pairs that enter exactly seven vertices. Finally, in step four he kept only the pairs that entered all seven vertices at least once. Having done this, he amplified the result and, by doing sequencing, determined the correct solution.

Although biological computers are much slower than electronic computers, their overall increase in speed is enormous because biological computers simultaneously test all possible solutions. There are other advantages. A biological computer is a billion times more energy efficient, and it has the potential to store information in just one-trillionth the space of conventional electronics. But more important is the parallelism. A DNA computer can have an unimaginable number of molecules performing this type of calculation simultaneously. According to Professor Richard Lipton of Princeton, a test tube could easily hold 10^{18} DNA strands, giving it the theoretical capability of performing one billion billion operations at once. So it's very well suited to performing massive computations such as breaking the government's data-encryption codes. It may take weeks or months to reach the solution, but the problem may not otherwise be solved. With this approach, we could solve many interesting and important types of combinatorial problems.

DNA could also turn out to be an extraordinary storage medium, because it should be possible to encode DNA sequences, store the DNA and then retrieve the data, which would be identified by a key word. We could search for the key word by adding a DNA strand whose sequence would stick to the word's DNA. The search would be performed completely in parallel, and that means memories could not only be huge and inexpensive, but they could be content-addressable or associative—that is, they would work in much the same way that some people think our brains do. Just to give you a feeling for the numbers here, if you had one pound of DNA molecules suspended in a cubic-yard-sized tank—about 1,000 quarts of fluid—you could create a memory bank that would have more storage capacity than all the memories of all the computers that have ever been made.

There are numerous obstacles to doing this, one of them being accuracy. The DNA sequences usually do what they're supposed to do, but not all the time. In the polymerase chain reaction, for example, the DNA makes a mistake in one in 10,000 to one in 100,000 molecules. This means we need to understand how to do error detection and correction and what that means in a system with so many errors. The processing time is too long and you have to be a biologist to understand the steps. We need more rapid and more automated processes. Experts assure me that's coming—they just neglect to say when.

But I do want to point out that this is very similar to the ENIAC situation. Remember the room-size computer with technicians and engineers who are manually transferring the results among the accumulators? Well, now we've replaced that with a room-size computer that has technicians and biologists manually transporting the intermediate results—except this time, the results are in test tubes and petri dishes. But just as with the ENIAC, we had better not discount the possibility of a truly disruptive technology appearing in the next 50 years. I don't, but I think that few believe that this is the way to general-purpose computing. However, DNA computers could be used to solve certain problems or could work with hybrids of electronic computers. It's hard to disregard the massive parallelism that DNA computers make possible. Fifty years seems a reasonable amount of time, and as far as I can see, there are no basic limitations of physics in sight.

Now onto the last of the three alternatives: Optical computing. This is not a new research area; people have been pursuing it for over 20 years. The appeal for computing, of course, is that photons behave very differently than electrons, and their unique properties allow for fast communication and coherent interaction between them. We've already exploited photons for communication data storage, audio CDs, video discs in the 1980s, the current IRDA (Infrared Data Association) devices, and the fibers and lasers that are forming the backbone of the new information society. So interconnection networks at the heart of future electronic computers almost certainly are going to be optical.

But what I'd like to explore is the idea of an optical computer, not its use as an interconnection device. Across the globe, researchers are pursuing different ways of implementing optical computing. It's possible, I think, to group these activities into two categories: hybrid electronic–optical computing, and all-optical computers. In the former, conventional electronic machines perform the general-purpose computations and control the resources. But the optical analog computers are included in the system to perform specialized functions. These exploit the key advantages of photons: massive parallel interconnection, high-speed processing and low-power data transmission.

An example of one such optical hybrid is the Fourier transform processor. A lot of these exist today, and the Fourier transform is the heart of many optical computing approaches. Here's how it works: Basically, a monochromatic laser light, split into a parallel beam, passes through a spatial light modulator driven by an electronic computer. This modulator presents the input image to an optical system that creates—in parallel and at the speed of light—the Fourier transform of the image in both amplitude and phase. This has to do with the very elegant characteristic of a lens to actually create the Fourier transform if we choose the focal lens properly. This can result in an extremely fast processor, whose computation time is independent of the size of the input and output patterns—another one of those exponential advantages that we find so interesting.

The second category—an all-optical approach—attempts to use only optical components to perform all the functions of the computer, right down to the transistor. Optical switching can occur at subpicosecond rates, maybe a factor of 1,000-times faster than the nanosecond rates available with electronic switches. But to me, the all-optical computer seems very distant, even though we're starting to see lasers that are fast enough to operate in this domain.

There are a great many serious challenges. Perhaps the most important one is that we don't yet have a practical optical memory. We have delay lines and holographic stores. But in order to make such machines operate the way electronic computers do, and in order to avoid going back and forth and in and out of electronic memories, we'll have to develop suitable electronic memories and figure out what optical VLSI looks like so we can make things short enough. Even if we're able to surmount those difficulties, we'll have to figure out how to program such machines. The I/O problems and all other problems have to do with packaging. But if we could solve them, what a wonderful technology we would have for building neural computers, doing image processing, pattern recognition and large-scale simulation.

By necessity I summarized these three computing alternatives superficially and incompletely, and I apologize for this to many of the people here who've been working on them. I want to add that these are not the only promising approaches. There are quite a few others. My real aim in this talk has been to create a reasonable doubt in all your minds that 50 years from now, the world of computing will simply consist of people using Decium Pros running *Windows 47*. There are a lot of obstacles to overcome, but 50 years—especially with the World Wide Web and the emerging infrastructure at our disposal—is a very long time. When I stop to analyze these three alternatives, the conclusion I reach is that in the future we're going to communicate with photons, but we're going to compute with electrons. I feel this way because of the tremendous disparity in the coupling forces between charged particles and photons. It seems to me that computing, which involves changes of state and switching, is best done with strong forces such as the coulomb forces among electrons. Just the opposite will continue to be true for communications, in which the far weaker photon-to-photon interactions present a decided advantage. If I had to place a bet, I would wager on some form of quantum computing becoming the way we keep the CMOS engine going as the widespread general-purpose computing technology. I think optical and biological computing are likely to have important, but more specialized, functions—usually in hybrid systems. Of course, there's a sheep named Dolly bleating in Scotland today, which almost all the researchers in the world thought would be impossible. So it may very well be that something new will come along that will make my prediction quite radical.

I'd like to spend a few moments speculating on some of the possible consequences of these unimaginably powerful computers and vast memories. I'm going to follow my own advice about not being too shy. Over the next half century, we're going to know things that we don't know today. For example, in 50 years we may really understand how the brain works. We'll map it, we'll understand the interconnections, we'll understand how our sensory organs are connected and how we really do this miracle of perception. So the question is, if you imagine that we could learn those things, then would we have the computing power to build an auxiliary brain, perhaps a wearable one? A brain that could augment our abilities to reason, to remember, to communicate—the kinds of things that separate humans from beasts? Could we use these computers to dramatically change the way we think about things? For example, many people have speculated that such a device might have the capacity to store all your life's experiences. That is, you'd wear a bunch of sensors, and whenever you turned them on they would record your complete sensory experience. I kind of like that idea because it would let you experience everything possible in, say, the first 40 years of your life, and then like the romantic poets, you could spend the rest of your life just sitting around reliving all those things you had done when you were young enough to do them.

But imagine also that we learn a lot about language and speech recognition. With some commensurate breakthroughs, why wouldn't it be possible to have a brain that could do language translation, perhaps in real time? Let's stretch credulity to the absolute limit and say that this knowledge of brain function and our ability to map them individually brings us to the point where we can understand what distinguishes one person's brain or sensory apparatus from another. Then you might be able to create a transfer function—I have a feeling this will take more than 50 years—which would impose the characteristics of one brain onto another. When you went to the opera, you would hear it through the ears of the diva; when you sampled wine, it would be the sommelier's tastebuds that would create your experience. We're going to know a lot more about how our bodies and our emotions work. Could we then make a computer with a sense of humor?

Let me conclude by paraphrasing something George Bernard Shaw once said: The reasonable man seeks to understand the world, and then adapt to it. The unreasonable man seeks to understand himself and try to change the world so that it matches what he wants it to be. All progress depends upon unreasonable men, and I hope that some of you will be provoked to start thinking quite unreasonably. The underlying question is this one: What will it be like to be a human in the year 2047?

PATTIE MAES

Very Personal Computers

4

James Burke

Apart from anything else, our next speaker is an amazingly productive person—she has a list of books and papers as long as your arm. She began life, so to speak, at the Free University in Brussels, Belgium, where after a number of years of reflection she got her doctorate in Computational Reflection. In 1989, she took a visiting professorship at MIT and never really went home. You can tell how serious she was about it—Brussels being the food capital of the world. In 1991, she accepted a faculty position at MIT Media Lab where she's now associate professor. She's also Sony Corporation Career Development Professor of Media, Arts, and Sciences. So far she's worked on intelligent office systems, problem-solving strategies, object-oriented languages and computational reflection, and her most recent work is on a theory of action selection and learning in an autonomous agent using a distributed model. In 1995, she founded Firefly Network, Inc. in Boston—one of the first companies to commercialize software-agent technology and to bring collaborative filtering technologies to community-building on the Web. The name of her game is electronic agents, what might be called the significant other of the 21st century. She is here to speak about that and other exciting, unusual and sometimes disturbing things. On the topic of how "personal" personal computing is going to get, please welcome Pattie Maes.

Pattie Maes

I don't know why we have such a fascination with anniversaries. Personally, I get less and less excited every year about celebrating my own. We seem to have a special fascination with anniversaries with lots of zeros in them, like 10, 20, 50. It would seem to make more sense for the ACM organization to celebrate the 32nd or the 64th anniversary rather than the 50th. Then, at least, computers would have some chance of understanding what we're talking about. But more seriously, I think this is a wonderful opportunity to reflect on how far we've come, where we're going and whether we're really on the right track. Since my background is actually in artificial intelligence, I see this as an opportunity to reflect about what AI has done in the last 50 years. AI is as old as computer science and the first computer scientists like Turing were really the first AI researchers.

The goal of AI is to build intelligent machines, and there's actually a dual justification for this research. On one hand, we hope to clarify or to get some insights into human intelligence by synthesizing computational forms of intelligence; on the other hand, we hope there will be a lot of practical applications for these intelligent machines. AI's Holy Grail is a computer that is like a human, that is as intelligent as, for example, C3PO, or HAL or lots of other examples from literature and science fiction movies. Some of you may be familiar with the Cog Project at MIT, where Robotics Professor Rodney Brooks is, in 10 years, trying to build a robot two-year-old. The head of the robot has exactly the same degrees of freedom, the same sensors such as cameras for eyes and microphones for ears—everything that humans have. The goal is to build a computer that can do what a two-year-old can. There are lots of other projects like it.

If I reflect back and look at what we've achieved in the last 50 years, then I'm actually very disappointed. We haven't really gained that many insights into human intelligence by building artificial intelligence machines, and there aren't that many practical applications we can point to. There really isn't a big AI industry today. So I wonder whether we've taken the wrong path somewhere and whether the solution may be as simple as reversing the acronym and working on what I call IA, which stands for intelligence augmentation. What I would like to argue is that we should try to work toward a future where we don't try to build these stand-alone intelligent machines, because there are far more pleasant and easier ways of reproducing human intelligence. Instead, we should be focusing on building combined forms of human and machine that are superintelligent and where both complement each other.

So what I'm really talking about is prosthetics. As people, we have a history of inventing prosthetic devices such as glasses, hearing aids, voice synthesizers and cars. All of you Californians would be complete invalids without your cars and bicycles. These devices have become part of ourselves, and we really can't live without them. I'm arguing that we should be working on prosthetics for the mind. Why? Personally, I believe that there's a whole set of problems or "bugs" in our minds. For one, we have lousy memory. We forget the names of people we've met, what they looked like, where we saved a file, where we left our keys or our glasses. Imagine how much more efficient the world would be if we had better memory. Another limitation is that we're only good at doing one thing at a time. We're not good at multitasking or simultaneously dealing with many different things. Another limitation is that we can only be in one place at one time. We also have cognitive weaknesses: We're very bad at logical reasoning and at dealing with probabilities—otherwise national lotteries wouldn't exist, or they wouldn't be as popular. We're also very slow to process large amounts of information. I have a lot of respect for Mother Nature and for her ability to evolve really good solutions to every problem in every environment. But currently there is a mismatch between our cognitive abilities and the environments we live in. This is especially the case for the digital world. Maybe nature and evolution haven't had time to catch up and transform us into an entity that is better adapted to dealing with this complex world.

There are so many things to keep track of every day that we all have to learn more and remember more. So if we think about resolving this problem of information overload by building combined forms of human and machine, then it becomes important to think about what the human mind should be used for versus what prosthetics should be doing. Computers and people are really good at very different things. People are good at things like aesthetic judgment, understanding things, reasoning, problem solving, being creative, etc., while computers still aren't good at these things at all. Computers, on the other hand, are good at a set of things that people aren't good at, like remembering a lot of facts, searching through large amounts of information, being in many places at once and doing many things at one time. For the last five years, since I joined the MIT Media Laboratory, I've undertaken this intelligence-augmentation research project and we've built the first prototypes of combined human–computer systems that could be said to be superintelligent. Some of these devices deal with memory augmentation—they augment limited memory—others function as extra eyes or extra ears; others automate some behavioral patterns to save people time to do other things; yet others help deal with information overload or may help with matchmaking—finding other people who share certain interests.

Finally, we've done a lot of work on computer entities that help deal with transactions of goods. I'm waiting for wearable computers to get a bit smaller and more fashionable, but all of my students these days walk around with them all day long. These computers have a lot of sensors that can tell what a particular student is doing. For example, they have a GPS (Global Positioning System) to keep track of where the student is going and software that continuously remembers everything they do, every place they go, and who they talk to. It keeps track of what applications they're using at what times, who they get E-mail from, send E-mail to—everything. The machine basically acts as an augmented memory, an external memory. You can ask it to help you remember a certain person, a certain place, a certain action that you took, where you stored a certain file, who you met in the hallway on your way to the bathroom last Tuesday. You can give the computer some contextual information about what you're looking for, and the system can help you retrieve it.

The system is also used to provide just-in-time information, proactively. Just like your own memory, the augmented memory continuously reminds you of things related to your current situation and your current environment. For example, if one of my students is walking past the library in Boston, the little monitor in front of his eye may give him some information about the current exhibits going on in the library. When I talk to him, he automatically sees all the recent E-mail messages I've sent. It's a little spooky sometimes because these students seem a lot smarter than they really are as a result of having access to all the right information at the right time. The device has a window at the bottom of the screen in front of their eye that continuously gives them reminders of information that is related to their current situation. This involves both their physical situation as well as what they happen to be doing with the computer—like E-mail, for example.

Another area we've explored is the use of computers to provide extra eyes and ears that monitor the things that people really care about—not just bits or events in the digital world, but also events in the physical world. We've been building little software entities that you can create and leave behind at a certain database that continually watch things that interest you and tell you when changes occur. For example, they'll let you know when your stocks have suddenly fallen or when a certain Web site you care about changes. It's as if you had an eye or an ear placed on that particular database. We're also doing this in the world of atoms by embedding sensors and computing devices in your fridge, for example, so it can monitor how much milk you have and remind you to get some if you're running low; or placing a little sensor in the coffee machine at your office that tells you whether there's fresh coffee or whether it's old—in essence, warning you to stay away if you dislike making a new pot of coffee.

Another area we've explored is the automation of behavioral patterns. One very early system we've built has a little face representing the agent. This eager assistant continuously watches you deal with a particular software application, say, a meeting-scheduling program not unlike *Meeting Maker.* Basically, that little agent will pick up some of the patterns in your behavior and will then suggest automating those patterns on your behalf. So, for example, my agent may notice that I don't like to have meetings before 9:30 in the morning. It may pick up on regularities such as which meeting rooms I like to hold them in or how long I want them to last. After a while, this agent can offer to take over some of my meeting scheduling because it has learned the way I like to schedule them.

Another class of IA systems help filter information in asynchronous as well as synchronous ways. Synchronous filtering employs agents acting as online guides that continuously point you in certain directions they think you will be interested in. For example, we've built a Web browser that continuously watches what Web pages you go to, analyzes them and builds a user profile—a profile of your interests. It may learn that you're interested in classical music, that you're interested in certain key words or that you're interested in Web sites that contain lots of pictures. The agent acts as an advance scout, so whenever you are at a certain Web page, it looks ahead and follows all the link's pages up to a certain depth. Then it checks whether any of the pages in the local neighborhood have any of the characteristics that you seem to be looking for and points you in that direction. Asynchronous filters work about once every day to check whether there's new information that may be of interest to you—again based on the system's sense of your interests. Take *Firefly* technology, for example, which helps you find not just Web sites that may interest you, but also movies, music, and soon books, software and a whole range of other categories of information. Again, it works by building up a profile of your interests and proactively suggesting things it thinks you might be interested in as well as when you ask for it. So, for example, my agent knows that I like Woody Allen movies and when a new film of his comes out, the system will tell me about it. I don't have to ask it every week, "Are there any new movies I would be interested in?" My personal-information filtering agents are looking on my behalf.

One of the hardest problems people struggle with, although we never really acknowledge it, is matchmaking. Finding a mate—or just someone to talk to—is very hard. We may spend 10 years communicating with another person, finding out what they're like, finding out whether they would be a good match, and then spend another 10 years learning that it was the wrong mate after all. This is really a more general problem of finding other people who potentially share our interests. We have been building matchmaking software, not just for romantic matchmaking, but for solving this broader problem. In this model everybody has an agent who watches your E-mail, your Web browsing, whatever you want it to look at, and extracts some of the interests you have. For example, the agent may learn that you're interested in scuba diving, or classical music or whatever. Then it will talk to other agents and take the initiative to introduce you to people who share your interests. The same system is used when you meet someone to create a link and tell that person about some of the interests both of you have in common—at least the stuff you want to reveal during the first meeting.

The last set of intelligence augmentation software entities we've been working on are agents that engage in transactions on your behalf. We have a Web site called Kasbah, currently accessible only to MIT people, but we soon hope to make it available to the world at large. At Kasbah site, you can create agents—software entities—that will buy or sell something on your behalf. In reality, this is a very inefficient, time-consuming process. We built a system where if you want to sell a second-hand computer, for example, you spend only a minute to create an agent just by filling out a form. You tell it what computer you're selling, the deadline for its sale, the starting price you want to ask, the minimum price you're willing to accept; how that agent should negotiate on your behalf, whether it should be a tough or flexible bargainer, whether it should be selling only to local people, what its level of autonomy is and how often it should check back with you and tell you what has been happening. Once you create an agent like that, you can go off and do other things while it negotiates with agents of potential buyers. It lets you know when it has found a deal, or if it thinks it won't find a deal in the available time. You can have as many agents as you want continuously acting on your behalf.

It gets even more exciting if you have all these software agents collaborating with each other. For example, if the agents monitoring the level of milk in my fridge could talk to the remembrance agents on my wearable computer, then when I'm driving past a grocery store, it could display a message on my little screen saying, "You should run in and get some milk because you're running low." If my shopping agents talk to my matchmaking agents, then I could create an agent to buy a Saab 900, for example, that could actually look up other people trying to find a good deal for that same car. My shopping agent could even hook up with these other agents and create a little Saab 900 consumer cartel. They could go to the Saab dealership and say, "We'll buy 20 of them, but we need to get them at this price." If my eager meeting-scheduling agent knows that I have a meeting with, say, IBM people, it could tell my filtering agents that I want to see some more stories on IBM, because I want to appear as if I'm really interested and that I'm following what they're up to.

So what hardware and software will be involved in realizing this picture of intelligence augmentation? In terms of hardware, one key component will be wearable computers—they'll get much less clunky and they'll become part of your glasses and clothes. If you are going to have your intelligence augmented, it's important to have this with you at all times—you don't want to leave half your brain at home, so to speak. Ideally, they would be implanted so you couldn't forget it, but until that is possible we'll have to integrate it with the other stuff we wear on a daily basis.

Another key hardware component will be what we call at the Media Lab "things that think." These are everyday objects like toys, toaster ovens and doorknobs, which have embedded sensors, processors and communication devices. If we're going to build these agents—the software that knows what you're up to and can help you achieve your goals—it's important that these systems operate not just in the digital world, where things are easy to sense, but also in the physical world.

In terms of software, a key component will be software agents. What we mean by software agents is really a new way to think about software. Unlike today's software, it is very personalized. It knows you, it knows your habits and your preferences. It's very proactive; it doesn't just sit there and wait until you tell it to do something, but it actually tries to do things on your behalf and achieve some of your goals. It can act autonomously, is very long-lived and is continuously running. Finally it is continuously adapting to your changing interests.

Another key component on the software side are what we call "digital ecologies"—massive collections of machines as well as people that together can solve certain problems or perform certain activities in a radically distributed way. These kinds of systems are typically very adaptive; none of the components is critical and they're very robust such that machines or people can be removed and the system will still be able to perform. Many of the laws that apply in nature will apply in these software ecologies; things like competition, natural selection and evolution. We'll see more and more instances of small efforts by many people as well as machines to solve certain problems and deal with certain issues rather than through large efforts by very few. This will result in a way of doing things that is more efficient, more adaptive and more robust.

I would like to conclude by talking about some very important design challenges or user interface (UI) challenges that we should keep in mind as we build these devices. First, people will adopt these devices only if they can really trust their computer; and trust in this case means understanding what the computer can and cannot do and to some extent how it operates. Second, humans always have to feel in control—you don't want this computer running your life. Last, and most important, is privacy, because once everybody has these devices that continuously keep track of everything you do, everything you read and everywhere you go, it becomes really important that the collected information is safeguarded and that you are the only one who decides what happens—or doesn't happen—with that information.

NATHAN MYHRVOLD

I, Software

5

James Burke

Our next speaker takes very little time off. Well, ever since computers and E-mail made our lives so much easier, who does? But here we are. Anyway, when this gentlemen does take time off, it's to do things like work as an assistant chef in a leading French restaurant in the Northwest. He also has competed twice in the World Barbecue Championship in Memphis, Tennessee, and won first and second prizes—not at the same time—so this guy is really cooking. Also, his education includes certificates in mountain climbing, formula car racing and photography. He has degrees in math, geophysics, space physics and mathematical economics as well as a doctorate in Theoretical and Mathematical Physics from Princeton. As a post doc, he was in the Applied Math and Theoretical Physics Department at Cambridge, where he worked with Stephen Hawking on cosmology, quantum field theory, curved space time and quantum theories of gravitation. Fortunately, he also speaks English.

When he returned to the U.S., he founded Dynamical Systems Research, Inc. and sold it to Microsoft. He then joined Microsoft as director of special projects in 1986. From that he became vice president of Microsoft's advanced technology division, responsible for product development—stuff like interactive TV, advanced graphics and new forms of consumer computing. Today, he's chief technology officer and is responsible for really serious amounts of company revenue. He sits on Princeton's Board of Trustees of its Institute for Advanced Studies as well as on the physics department advisory board. He is also a member of the National Information Infrastructure Advisory Council. So, he's one of those few people who really can say, "Been there, done that," and he's only 36, so listen up. Please welcome Nathan Myhrvold.

Nathan Myhrvold

It's a pleasure to be here. Not only is this a very distinguished audience in many respects, but my mother is sitting near the front row, so if you know my speeches, you'll know this one's constrained a bit as a result. But what the hell—oops, I shouldn't have said that. The topic I agreed to talk about was the next 50 years of software. But after thinking about it, I decided there was a better title: "Software: The Crisis Continues." I'm going to talk about this more from the perspective of being an unrepentant programmer and theoretical physicist than from any other perspective.

James did a brilliant job of summarizing the history of western civilization. I thought I might do the same for the universe, because that's what cosmologists are all about. The universe started as a quantum gravitational fluctuation that started expanding exponentially. It continued to grow at an enormous rate, which caused the hot quark plasma to cool, nucleating elementary particles. Radiation from this fireball was red-shifted down to 3 degrees Kelvin—you can still see it with a satellite—and matter ultimately condensed into the San Jose Convention Center. As you can see, I've skipped over a few little points.

Now, let's talk briefly about the history of information, since that's very topical for a discussion of software. Information didn't really exist per se. Until writing existed, we couldn't really pass information very effectively. The best ideas of our species were literally lost to the wind. So writing was a big deal. You could say it was one of the first information revolutions. But carving on a stone tablet or on some of the other mediums, like calligraphy on parchment, didn't reach enough people for writing to be a big deal—that is until Johann Gutenberg invented the printing press with its movable type.

That started an incredible information revolution. Since then, many incremental improvements were made, but it wasn't until the computer was invented that mankind had a fundamentally new way to store, manage and manipulate information. Like that first writing, though, it didn't reach many people. It was confined to the raised-floor temples where the great mainframe was stored. Well, the microprocessor changed that by bringing computing to everyone, and now networks are hooking us all together. The important thing about the Internet isn't that it's hooking computers together; it's hooking people together, allowing them to communicate. The combination of cheap, ubiquitous information distributed this way is causing another information revolution that will have an impact on our society at least as great as the Industrial Revolution did.

The fundamental technical driver for this is Moore's law. Roughly speaking, in the last 25 years we've had an increase by a factor of a million in the processing power of computers. If this continues, in the next 20 years we'll have another factor of a million and there's reason to believe it will go on for another 40 years. Now, this is a long-term extrapolation, but I'm happy to be making it for three reasons: First, I'm a physicist and the fundamental physics are well understood. As Gordon mentioned, a team at the University of Minnesota has built a RAM cell that is seven nanometers on the side, using a quantum dot. Second, I'm a software guy, and this is a hardware problem. When you start tossing around factors of a million it's very tough to figure out what it really means. One way to get a grasp on this is to say that a factor of a million would shrink a year into 30 seconds. That means in 30 years, computers will be able to do in 30 seconds what one of today's computers of comparable price would take a year to do. But 40 years hence, these computers will do in 30 seconds what one of today's computers would take a million years to do. Those gigantic kinds of numbers boggle the mind. It also explains how a cosmologist can find employment in this industry.

What will we do with all that power? The answer is software. Software is the process of taking all those bits and bytes and cycles, and hopefully turning them into something useful for people. Everybody has their own laws—Metcalfe's laws, Bell's laws and Moore's laws. Well, here are Nathan's laws of software:

Nathan's first law of software: Software is a gas; it expands to fit the container it's in. (Thank God those guys that make containers keep making them bigger and bigger and bigger.) Here's an example of this: Microsoft *Word* started out as a program with 27,000 lines of code. That was back in 1984. Currently, *Word* is at 1.7 million lines of code, which isn't really fair, because it also draws on a shared program library of about another million lines. Lest you think this is a Microsoft phenomenon, I've taken the size of a very popular Web browser that has a doubling time of 216 days; it's growing at the rate of 221 percent per year. There were no Net browsers a few years ago, and we're just starting to learn what they're good for. We're adding more features all the time, so it's going to keep growing and growing.

This takes us to Nathan's second law of software: Software grows until it becomes limited by Moore's law. If you expand into a vacuum, you quickly fill the size of the container you're in. This is what we're seeing with browser software; it continues to expand in features and functionality. Eventually, the hardware does impose a limit, so software fundamentally grows faster than Moore's law. When a new machine comes out it has a bunch of headroom. Boom! The software grows and bumps into it. This is because software is able to bring any processor to its knees just before the new one comes out.

This takes us to Nathan's third law of software: Software growth is what makes Moore's law possible. That's why people buy the stuff. If we hadn't brought your old machine to its knees, what's the point of buying the faster new one? It's been bugging me for years, but I have great respect for the tremendous things these hardware guys do in doubling those transistor counts, making computers faster, cheaper and more powerful. But guess what? They couldn't sell a damn bit if software didn't make that stuff useful. That's why chips keep getting faster. It's also why chips tend to get faster at the same price rather than getting cheaper. Of course, some chips do get cheaper, and you wind up getting microprocessors in your watch, toaster oven and other things. But if they only got cheaper and people didn't care about absolute performance, the hardware makers wouldn't be pushing the edge of technology quite so much. I believe this is going to continue as long as there are people writing software—even after people have stopped writing it.

That brings us to Nathan's fourth law of software: Software is limited only by human ambition and expectation, which is to say it's hardly limited at all. It is fundamentally impossible to satisfy people—I have no proof of this, but it's my empirical observation—at least at this stage of technology. Maybe someday every piece of software will work perfectly. As long as there is the opportunity to create new algorithms, there is the possibility of developing new applications. In the last couple of decades, graphical computing became possible, which gave birth to desktop publishing along with visualization and CAD. This all came about because computers got powerful enough and the right technology was created for graphical tasks. Suddenly there were new applications. People who never thought they needed a computer before, needed them now. As the price continues to drop—and it will—new users will be able to afford it.

Finally, the ever-changing notion of what's cool produces programs that really push the state of the art. These aren't things like *Windows NT* or Microsoft applications. The software that really takes maximum advantage of the new hardware is games. A lot of the ones that are going to come out in the next year are going to require at least a 120 megahertz Pentium and 16 megs of RAM, and a lot of them will have features that will only work with MMX. Why on earth would people selling a $49 program require such a powerful machine? Because people think it's cool, and cool is a powerful reason to spend the money.

There may be a fly in the ointment, though—the software crisis. If you look at any collection of articles about software—it doesn't matter if it's from the 1950s or very recent—somebody is always talking about the software crisis. Software has been in a constant state of crisis, which centered on how to develop high-level languages and all the necessary software. If you read any one of these articles, they pose the question of whether there is some fundamental limit to complexity that will stop us from making software more functional. We're just going to hit a dead end. Other people ask another way. They say, "Well, you know, civil engineering is a very predictable field, and electrical engineering is a fairly predictable field. How come software engineering is so immature? Isn't it going to grow up and be like one of these real branches of engineering? Is this crisis ever going to end?" Well, of course not. Software is always in crisis, and most of this discussion stems from a fundamental misunderstanding of what software is about. In fact, I think it's going to be a perpetual crisis, and I confidently predict that someone will talk about software crisis jokes at the ACM meeting in 2047.

People have touted lots of panacea solutions to the software crisis: high-level languages, object-oriented programming, component software. There would always be some new thing to set us free. But all those things *did* happen. We *do* use high-level languages and object-oriented programming; people *are* structuring applications as components. What happens? Every single bit of that progress is instantly absorbed—a victim of ambition, expectation and creativity. When a software engineer has a great idea, he tries to do it. So it's a naive notion that just over the horizon lies a miracle technology that will make software problems disappear. That would be possible only if we kept writing the same kinds of programs.

The trouble is, we don't. We all write ever-larger ones. The moment it becomes possible to easily go distance x, a bunch of crazies make it possible to go 2x. They'll always be at the leading, bleeding edge. In fact, software productivity is up dramatically. In the last decade, productivity at Microsoft is up by at least a factor of 10. That's how we're able to produce the programs we do. We reuse components (reuse is a big deal for us), we use high-level languages and object-oriented programming—every trick in the book. But we still push the limit, and so does everybody else in this industry.

The fact is, software is never going to be easy until we let up on the requirements. People sometimes ask me whether it would be better to have a piece of software that didn't have lots of new elements, one that kept only the functionality but just did a better job of it. That probably would be a great thing for some kinds of software. When I press the brake pedal in my car, it's software that stops it—the anti-lock braking system. I'd love to think that software is really stable, but when it comes to making the next version of *Doom* or *Myst,* people are going to buy what's cool—everything else is going to fall somewhere in the middle. However much we make it easier, somebody is going to push the boundary.

There are all sorts of things we'll be able to do with this new software. My friend Steven Spielberg made a great movie a couple of years ago called *Jurassic Park.* I submit that if we can bring back T-Rex, the king of dinosaurs, we can bring back Elvis; and if we can bring back Elvis, we're going to bring back all kinds of people by creating entirely synthetic actors. If you thought people from L.A. were fake today, just wait. But you have to remember that the technology Steven Spielberg uses now will become available to everybody within a few years. In five years a child's toy will have the same power as the machines used to make *Jurassic Park*—maybe even next year. You really have to look ahead far enough to beat this exponential wave of computing power.

As software technology progresses, we're going to see a variety of new techniques—new panaceas, if you will—that will be both used and abused. The first is genetic programming. This is the notion that you evolve programs to do what I call "software husbandry." Rather than going out to tend the flocks and making sure you're breeding your cattle or sheep effectively, you'll be tending their programs and using very sophisticated simulations of evolution to create software. This has already been used in some small areas and has lent itself to a variety of research.

But genetic programming begs an interesting question about the complexity of software versus the complexity of, say, the genome. Let's do a little test here to see which is bigger—*Evita,* the motion picture staring Madonna, or Madonna the genotype? This is an odd comparison, but which one is larger in terms of information content? It's the movie with four gigabytes; Madonna's

DNA is around one gigabyte, and that's not even counting the fact that 98 percent of it is shared with chimpanzees. We can go further and look into the genetic complexity of human beings. If you took all the humans on Earth and put their genomes together on one big server, they would occupy about 3.7 terabytes. We can compress the relatives, so it turns out that all the DNA you have, except for a couple of random mutations, came from mom and dad. So if we use mom's and dad's to get yours, all we have to do is record which DNA mix you got. That should take you down to less than a terabyte—about the size of a big Web site. If we did the same for all the animals on Earth, their cumulative genomes would be about the size of the Web. This isn't to diminish the importance of biology, but it is stunning to think that the process of evolution over a period of three billion years has created a store of information encoded in DNA in all the organisms on this planet, and in just a few years we've created a store of information of comparable size—at least within a factor of five or 10. That's a sobering thought. It doesn't make the Web any more interesting; in fact, a lot of stuff on the Web is pretty poor. (Of course, a lot of that DNA is encoding for slime mold, so I think it evens out.) But these examples show that although genetic molecular biology and genetic technology are tremendous things, computationally they're not that complex.

Let's turn our attention to the ultimate computer—the human brain. Every day, we learn a little more about it. There was a period of great optimism in the 1960s and 1970s when people thought artificial intelligence would equal or surpass the human brain despite the lack of understanding of how it worked. Orville and Wilbur Wright built an airplane without knowing how birds flew, so why couldn't we make programs that are smart? But we haven't, and it has caused people to be very skeptical. I'm shocked that this audience voted that there won't be artificial intelligence by 2047. I think there will. We might develop it by following Carver Mead's approach: understanding the neurons, understanding the brain and emulating them in hardware and software. We might do it some other way, but I'm convinced, and I think the first edges of this will appear well before 2047.

The downside of trying to equal a human brain is very daunting. My dog does most of the things people talk about when they refer to AI—speech recognition, the ability to do simple tasks, reasoning, vision—and today's computers aren't as smart as my dog. So, I'd like to set a goal: Within the next 15 years, we will have software that is as smart as my dog. If we can get it as smart as a dog, then it must be pretty easy to get it to the level of a pig or a baboon. Maybe that will take another 10 years, maybe only one, but if we can get that far, eventually we will have computers that are at least as intelligent as we are—in some ways, far more so. My laptop already multiplies a lot better than I do. We'll need those machines to perform only a very specific set of important tasks, so in 20 to 30 years, well before the 50-year time frame, I think AI will be here. If you accept this prediction, one thing to remember is that the brain has no Moore's law. We're not getting smarter every year—they are, and it's very easy to underestimate an exponential.

Now, suppose we had a computer that was as smart as a human. How would we program it? When you turn a computer on, you have to wait until it boots up before it can do any good. I've examined the same issue with humans and it turns out that the boot time is about 20 years from the time they're born to the time they actually get a useful job. It's a laborious process to try to get any programming into them. This presents an interesting problem: We may be able to make computers as smart as ourselves, but we won't be able to teach them; or they will learn for themselves, in which case they're going to make a hell of a lot of stupid mistakes. Just imagine the AI equivalent of a frat house.

There's a solution to this Moore's law brain issue, which is to scan the brain of a person and put it into the computer. There are two ways to do this: One is based on a wonderful project called *The Visual Man, The Visual Woman,* where dead humans are frozen and sliced very thin. Each slice is photographed to produce a three-dimensional reconstruction, an approach that is going to revolutionize the teaching of medicine. Right now, the slices are a third of a millimeter wide. Once that can be thinned to a few microns, it will be possible to section the brain that way and record it. The limitation is that this method would only show the brain's static connections; it wouldn't reveal all the dynamic functions that might be important. The second way to scan the brain is magnetic resonance imaging, or some new form of imaging to actually find out what's going on. In fact, a procedure called positron emission tomography actually lets you see what parts of the brain are active at any given time. You can show people pictures of tools and fruit, for example, and discover that the names of those tools are stored in one spot and the names of the fruit are stored in another.

Okay, suppose we do that? What sort of operating systems are we going to need for managing human software? We have to house all those uploads for people who decide to make that big leap to the matrix. They're going to need housing, which is a lot like an operating system function. Ultimately, I think we'll be able to treat humans as application programs; they might have their own server, they might live in a condominium server with a variety of others. They will also need a bunch of fundamental services and some peripherals to deal with the real world. My first job at Microsoft was development manager of *Windows 2.0,* which was released in 1987. Let's compare that to the features of *Windows* in the year 2047, when we're housing uploads. *Windows 2.0* had multitasking; *Windows 2047* will have multiple personalities, because after you've uploaded and you're doing one consciousness level at a time, why not start doing two things at once? You need multitasking. We had virtual memory in *2.0;* we'll need virtual memory in *2047.* The more you experience and cruise the Net, the more memories you're going to need to store, so we'll need virtual memory. *Windows 2.0* was compatible with DOS-based applications; *Windows 2047* will be compatible with meat-based humans. (Yes, there will be a backward compatibility problem with those meat-based humans for quite some time.) The big issue with *2.0* was whether we could break the 640-kilobyte memory limit; it will be 640 petabytes we're trying to get beyond in 2047. Of course, *Windows 2.0* had GUI (Graphical User Interface); I'm betting that *Windows 2047* has *you and I.* So I hope to address an audience much like this in the year 2047. I hope it's going to be a great experience for all of us. I hope I'm not talking about software. I hope I *am* software.

BRAN FERREN

The Lost Art of Storytelling

6

James Burke

Bran Ferren is somebody who lives in a fantasy world, but unlike the rest of us, he gets to live out his fantasies. Mind you, it wasn't always that way for him. He started out, like all good and serious people, at MIT. Then, at the age of 25, everything went wrong. He set up a company specializing in research and development, creative design engineering and systems for the visual and performing arts. So much for good education. At the same time, he started consulting for Walt Disney Imagineering. Then they bought his company, and in 1993 he joined them. By that time, he was an internationally acclaimed designer and technologist working in theater, film, special effects, architecture and the sciences. In 1995, he became executive vice president for creative technology, and R & D. What all this means is that if you've ever seen *Star Trek V* or heard of theme parks like the Twilight Zone Tower of Terror, Honey I Shrunk the Audience or the Indiana Jones Adventure, well, guess who? Of course, there is always his work on titles and special effects in such biggies as Ken Russell's *Altered States,* Frank Oz's *Little Shop of Horrors, Dirty Rotten Scoundrels* and many others. He has won prizes up the yin yang for everything from computer-controlled lighting to the advanced-concept optical printer to his own film, *Funny.* He's a member of the Motion Picture Academy of Arts and Sciences, and he's on the U.S. Army Science Board. That one I don't get, but never mind. One of the things he does a great deal of, however, is this. He's spoken to just about everybody over the last few years in the entertainment business and scientific communities. But having said that, I must say that when you're preparing for a conference involving him, he is the world's most elusive speaker. Although I think he's going to talk about the long-term impact of information technology on the entertainment industry and the whole business of future styles in storytelling, who knows? I just hope he's still here. Ladies and gentlemen, a warm welcome for Bran Ferren.

Bran Ferren

I get a strange feeling, because while I agree with much of what's been said here—and I think some of it is quite fascinating and provocative—I agree for completely different reasons, and it feels like I come from a different planet. So, you're going to get a talk from a fundamentally different way of looking at all of this. Take the World Wide Web, for example, and ask, "Why is it exploding?" Personally, while I think this Moore's law thing is cute and kind of handy, and that Internet interoperability standards were critical, the basic reason why I believe the Web is experiencing such phenomenal growth is simply because the computer is just getting good enough to be used for storytelling. And you're going to hear a storyteller's perspective because that's how I look at all of this. I eat, live and breathe story. To me, it's everything. I disagree with an observation made earlier today that the first technology to start all of this was writing, because I think storytelling significantly predated the written word. Whether you call it story, myth or fable, the art and science of storytelling is what I'm going to talk about relative to the next 50 years of computing.

I don't think that when we look back on today from far into the future that this will be thought of as the information age. The information age will be an interesting little curiosity—a blip. I believe that this era will be remembered as the storytelling age. If you look at computers as storytelling tools rather than as information-processing tools or basic communications tools, it becomes a lot clearer as to where they need to go. I don't mean to diminish interesting engineering aspects of our evolving computers such as the significant progress in processing power, memory or networking, because I do find that quite extraordinary. But every time I hear those things touted as being great contributions to our society, I ask myself, "Who cares?" What will they do to make our lives better, richer or more interesting?

I discovered the power of computers for storytelling a long time ago, because I've always lived and worked amid the arts, sciences and engineering. I've found that there are two kinds of people in the world: people who think there are two kinds of people, and people who don't. But beyond that, I think there really are two different sets of sensibilities, and they're often disconnected. These sensibilities are what drive both people and organizations in the process they use to produce new things. The first, for lack of another term, is the "requirements person." This approach is common in the engineering and technologies communities. You hear the word requirements a lot in much of industry and the military. The basic theme is: "Just tell me what the requirements are and I will deliver for you an engineered solution." The requirements process can be very useful. Sometimes people approach generating these requirements by following people around in their daily lives, observing them, doing focus groups, market research and a whole bunch of other things necessary to acquire enough history and understanding of the problem to develop an appropriate design and engineering solution.

So that's the first way of approaching new product development. The other way, which is more common among professional storytellers, I call the "big idea" process. It presumes that requirements are all well and good, but you're never going to do anything really new and innovative by asking people what they like or by paying attention to what they do. What we need instead is the big idea—something that's really cool and will basically change the way everybody thinks about a given product. Both of these approaches use intuition and complex professional skills. The requirements person often starts the process by doing market research and R & D, talking to the customer, holding focus groups, etc. This discipline is intended to result in a properly developed requirements document which becomes the ironclad specification for the product.

The big idea person's intuition is to approach the problem entirely differently. That person starts by thinking about how to design something unique that people will be passionate about. It's because he or she truly believes that they deeply understand this characteristic of what people want (often as contrasted to need) that gives them the confidence to guide a design without using requirements or even talking to the prospective customer. I'm not suggesting, by the way, that either a requirements or big-idea process is intrinsically superior. Both of these approaches can produce excellent results. However, certain tasks are much more suited to one approach than the other.

Try creating a movie using a requirements process or building a jet aircraft using only big ideas. For the right project, either—or both—can be very effective. In fact, some of the best organizations have learned to employ both, generally transitioning from a big idea to a requirements process as the project proceeds. This can be a bit tricky because the simple fact is that requirements people and big-idea people usually hate—or at minimum barely tolerate—each other! This presents a management challenge to build mutual respect for the others' methods and sensibilities. This is not easy when each group basically speaks in a different language. The words are familiar, but the meanings are often completely different. Getting these folks to communicate effectively is like trying to explain the emotional importance of the color red to someone who's colorblind. I say, "You know, red like blood," and they're thinking, "Yeah, gray like blood"; and "like red-hot steel," and they're thinking, "like gray-hot steel." We just don't have a particularly good language to bridge these worlds.

If you look back at our history from the perspective of the storyteller, you realize that every time a significant storytelling technology has been introduced to the world, it changed the course of civilization—or rather, people embracing these technologies changed the course of civilization. This includes language, writing, poetry, theater, printing, radio, movies, television, videotape, and now it's networked multimedia computers. The moment you introduce a decent new storytelling technology, human beings are so genetically wired with the compelling need to tell stories to one another that they frequently embrace it. Often these technologies change our world.

Now, I actually do believe that there are other types of needs people have in their lives beyond storytelling. On today's Internet we have non-storytelling interactions that are based on transactions, gaming and information transfer or retrieval. But if you take the position that a networked computer is a storytelling device, you'll do almost all of those things even better. Storytelling is such a compelling form of communication that if we figure out how to get computers to do this better, the positive impact they will have on our lives will be dramatic.

I am alarmed that many people look at the computers of today and think they're okay or, even worse, they believe they're actually good. They believe the hardware problem has been solved and now we're moving on to doing something about software. As soon as we get rid of those operating systems, everything will be fine. I am in violent agreement that our best software is primitive and in need of enormous improvement. However, I'm also here to tell you the hardware problem isn't solved and that the very best computers we have today are stupid, ugly and unresponsive. If you had friends like this, you would have a very lonely life. Yet somehow we accept them.

Why should computers be so ugly? Today's computers look and work like they were produced at the institute for the aesthetically challenged. They are the perfect example of a system that pays little, if any, attention to what human beings do or how we work. I guess the unwritten rule is: "You, the human, because you're smarter, will learn how to use the computer no matter how awkward that interaction might be."

I think it's possible for us to do better. When we look into the future, will the Internet have lasting importance or will it (as some have predicted) turn out to be a fad? On the famous Ferren scale of technological impact, we have citizens' band radio on one end, where a technology came and went in about 20 minutes. At the other extreme we find technologies that have served us well for a very long time, like fire (which works pretty much as well now as it did when man first stumbled across a burning bush and learned to tame it). The question is: Will future generations think of the Internet as being "CB radio-like" or "fire-like"?

Well, I am confident that what we're witnessing here is fire. The earliest technologies that helped turn Homo sapiens into a planetary leader were language and storytelling. A shared language was the key to effective human interoperability, and storytelling was the essential art of communications. Up until this recent cloning stuff (which, granted, could be really big!), the Web is probably the most important technology we've developed in our entire history since language and storytelling. The Internet may very well turn out to be more important than reading and writing, which very well may turn out to be a fad! Check back in about 250 years for an update.

So, I would argue that contrary to what you may have heard, storytelling is the world's oldest profession. It is an honorable and complex set of skills that benefits strongly from the talents and experience of professionals. I think it's really interesting when you start looking at the implications of some of our new interactive multimedia products from the perspective of the storyteller. The assumption is that everybody is as good at creating compelling interactive media as everybody else. The story (you should excuse the expression) continues that the Web will empower everybody equally as authors and publishers. Why should this be other than the fact that many people don't acknowledge that storytelling requires the skills of professionals? People don't assume this in medicine. If you have severe chest pains, you tend to go to a cardiac specialist. Few of us under such a health challenge would rush out to Home Depot to get a "Do-It-Yourself Coronary Bypass Kit." Odds are you would trust a professional. Yet when it comes to storytelling, people say, "Oh well, I'm just as qualified to tell a story as anyone else." It isn't true. If you were watching *Schindler's List* on television, and you had a little box at home with a knob that let you dial in "happy Nazis," "mean Nazis," "Nazis who speak Portuguese" or a whole collection of different options, it wouldn't make it a better film. The reason Steven Spielberg earns a great deal of money is because he is a master storyteller. And what master storytellers do is come up with methods to work around the limitations of the communications media they are given. That is a significant part of the job of being a professional storyteller. It's understanding the audience, how to touch them, how to reach them, how to speak to them and how to find ways to get around technical limitations. For the same reason that the path of technology can be quite predictable by looking at the trends and the physics governing it, so too can people.

People's needs have not changed a great deal over the years. To be successful at most businesses, you end up being in the business of adding value to people's lives. Measuring this kind of success is pretty straightforward. By and large, you're competing for customers' time and money. There's a myth that people have leisure time. Does anyone in this room have leisure time? And how many of you have spare money? The very notion that companies compete for people's leisure time and disposable income is fiction. Most people don't have any of either.

If I'm going to successfully compete and bring valuable things into your life, whether those things are network-computer based, pizza-based or whatever-based, I have to take your time and money away from other people who already have dibs on them. The importance we place on time and money changes as we age. When you're a kid, money seems like the big deal. In America, we teach our kids that money is really important starting when they are very young. But as we get older, time elevates in our consciousness. In fact, about 10 minutes before you drop dead, time takes on a whole new meaning!

So why is all of this networked-computer stuff happening now? Why is there this fire, this explosion, this exponential growth? After all, we've had the Internet for decades and until a few years ago, it remained a handy toy for the Department of Defense and academia. If you believe like I do that it involves storytelling, well, what's changing about that? Yeah, the technology is getting a bit better, but what's different about *us*? What's different about our communities? What's different about people? I think we're seeing a redefinition of the word "community" that extends our sense of place into cyberspace. We now have thriving, vibrant communities of people who have never physically met or even been in the same city. Yet they often share deep feelings and have established meaningful social relationships. To the best of my knowledge, this is unprecedented. It's like pen pals on steroids.

There's another emerging phenomenon: the generation of kids coming up these days who don't think computers are anything special or any big deal. They treat computers like they treat the toilet; it's always been there, has a useful and well-understood function, they learned how to use it at a very early age and it's now part of their regular lives. They are trained on both, and society expects them to understand the accepted use of both technologies.

I think another thing that's happened is that as a society—and when I say society, I mean the techno-American one that's represented in this room—we are entering the post post-Jetsons era. When I was a kid growing up in the '60s, my peers' vision of the future was the Jetson's. We suspected that our kids' kids would be like little Elroy zooming around on his levitating skateboard in a floating apartment house. Orbital Arcosanti. We shared the '50s notion that something synthetic (plastics) could be made to be superior to anything that grew in the ground or occurred naturally. Then we went through another period, which started in the '70s, where the vision of the future became a little house in the country with a lawn and a white picket fence. The Jetsons were passé; they were the older model of the future pushed asunder by the eco-green movement. This new model stressed the importance of sustainability, getting back to nature and the growing belief that people matter more than technology. Our image of the "ideal" domicile once again became a little house in the country with a lawn and a white picket fence. Recently, our sensibilities have once again started moving on. Now, we basically want the same little house in the country—but wired. We want a white picket fence that, in reality, disguises a phased-array antenna, communicating with the shiny new low-earth-orbit Internet satellites. Work and home have converged in cyberspace.

What's interesting is that we embrace these technologies because we think they provide solutions to the very problems we have created for ourselves with earlier technologies. Newtech fixes oldtech. In fact, we *are* starting to see some benefits—although modest. E-mail actually creates a way of interacting with people that is different from calling them on the phone or visiting them in person. The nonsynchronous nature of E-mail allows us to send and answer messages when we feel like it. Perhaps, in an idea suggested by writer and technologist Don Norman, it would now be nice to be able to talk directly to someone's answering machine—even when they're home. I don't want to talk to you, just give me your machine.

While much of this new communications technology is pretty spiffy, much is not. All too often the purveyors of new technology are providing really excellent answers—but to the wrong questions. This produces a whole variety of solutions to problems that nobody has. The real challenge seems to be knowing how to ask the right questions in the first place. I find it helpful to greet each new proposal from the perspective of "Who cares?" What will matter from the perspective of people like you and me, or the larger community that doesn't come to conferences like this? What's in it for them? Answering these kinds of questions and then effectively articulating these visions to others is a pure storytelling problem. When it comes to storytelling technology, however, we simply don't yet have a common language to talk about the science, or quantify the parameters of much of the communications technology developed over the past 50 years. We don't even have metrics for the storytelling resolution of human interfaces. By this I mean a means of quantifying how changing the technical parameters of a given multimedia communications system alters its effectiveness as a storytelling tool. In my experience, many of these relationships are both nonlinear and non-obvious.

I've been thinking recently about a concept I call the emotional resolution of a communications medium. For example, take a visual display system. The emotional resolution concerns how effectively it can be used to tell a story, as contrasted with the system's technical resolution (which refers to how much spatial, chromatic and luminance information it displays per unit time). While these attributes are often related, they are frequently different. Unlike the engineering perspective, which often holds that more resolution is better (until you can't afford it anymore, or you reach a point of diminishing returns), in storytelling, more resolution can hurt rather than help. In fact, the reason photographers use diffusion filters to weaken the technical performance of their camera systems is because it does a better job of storytelling. One trip to the cinema to see your favorite heartthrob captured with excessive resolution will prove this point. Experiencing an 11-foot-diameter nostril with hairs coming out of it, each and every one perfectly resolved, generally doesn't create a more meaningful relationship between you and the performer. In this example, adding technical resolution beyond a certain point takes you out of the story, rather than bring you in deeper.

Understanding these characteristics and learning how to manipulate them effectively is part of the technical craft of multimedia storytelling. It's a living. And it's how we can be dramatically more effective at communicating ideas in a rich and compelling manner. The body of technical know-how, when combined with the creative and aesthetic skills of those trained in the art, is the entire basis of the performing arts, much of the visual arts and, yes, even the mom-and-pop operations we call the entertainment industry.

I confess that I do get a little concerned when I start hearing a sort of anti-entertainment fascism starting to emerge—the idea that if something is entertaining, it is intrinsically inferior to that which is "serious." Earlier today, another speaker warned of the danger of education degenerating to "mere entertainment." In our schools, knowledge and the process of imparting knowledge often can't be done in an entertaining way because it's simply not considered appropriate. The fact is, we shouldn't have to apologize for things that are entertaining or fun. Fun is good, and forgive me, but I think it's just fine for kids to have fun and to be entertained while they are learning. What is it about the notion of being entertaining and capturing peoples' imagination that some educators find so frightening? Could it be that they are intimidated by the ability of good stories to hold peoples' attention?

This is not to excuse the enormous amount of junk we are exposed to on TV and in the movies. But this often sad state of affairs is a question of what stories entertainment executives choose to tell (and audiences choose to watch)—not the process of storytelling.

Storytellers have developed very interesting ways of working around the technical limitations of the media in which they choose to work. Let's look at the science of compression. Scientists and mathematicians have given us great insights into how to stuff 100 pounds of image into a five-pound bag. This technology is what has made fax machines, compact discs, networked multimedia and digital television practical. Ten years ago, in the dark ages of digital-compression technology, most professional communications engineers would have told you it's impossible to send a digital high-definition TV picture (with multichannel digital sound, no less) over a standard analog terrestrial broadcasting channel. Well, it turns out you can do this and more, but it takes more than just good mathematics. You need to understand something about the structure of images and how people perceive moving pictures. This viewer-centric insight led to techniques such as perceptual coding and semantic compression. These were among the breakthroughs that eventually led to the MPEG standards. Note the acronym stands for motion picture experts group. It came from people who make moving pictures (both kinds) for a living.

This is really neat, but storytellers have given us one of the most powerful forms of compression ever developed, one which harnesses the power of our imaginations to do the decompression. Storytelling compression lets the director of a horror movie create truly horrific apparitions guaranteed to make average viewers literally jump out of their seats. The master storyteller realizes that by putting his monster in the dark, he can tap into the inner fears that come from our shared life experiences. Remember the bogeyman in your closet as a kid? You didn't have to see him to be afraid of him. In the dark, there are no pixels at all, and that's what I would call pretty effective compression. The Stephen Kings of the world know precisely how to create vivid images that explode in your mind, often by using the power of words alone.

Many of the technical problems we need to solve to allow computers to fulfill more of their promise to benefit society are deep, complex problems. What I'm proposing is that we productively consider spending a little more of our time faking it. We all suspect that it's going to be really hard to make machines that are self-aware. I would guess that this will happen when we begin to reap the rewards of our children's children growing up with teraflop PCs in their homes. I predict that within 50 years, someone (aided by some*thing*) will create the first machine that will be self-aware. Sometime after this, its inventor (creator?) will realize that their machine became self-aware! And this will open up a new era of computation (and controversy). Perhaps self-aware storytelling machines will be the breakthrough necessary to power the educational applications of the future. But what do we do in the meantime? I would respectfully suggest that we devote a bit more of our attention to the fine art of faking it.

Faking is one of the cornerstones of literature, theater and film. It allows professional storytellers to be more effective without having to repeal the laws of physics or reincarnate dead actors every time they've been knocked off by fictitious bad guys. When you see a movie, the production team faked it. How often do you think when people get killed in movies, it's real? They don't do that very often in our business—at least not intentionally. We fake it. Somehow, faking it when it comes to computers and human interfaces is considered cheating. I guess this time I'm talking about "faking-it" fascism.

If our technologists forged (you should excuse the expression) a tighter bond with our storytellers and artists, I guarantee we would soon have a generation of computers that *seem* much smarter and more responsive to the needs of their owners. All it takes is the effective application of some smoke and mirrors, more sensitivity to how people interact with our world, and a thing called taste. It's okay to cheat if you're honest about it!

One essential ingredient to effective storytelling is the ability to establish a common context for discussion. To date, we have made little progress in making machines human-context smart. This is a big reason why they often seem so dumb. I maintain that the fact that they are dumb doesn't mean they necessarily have to act that way.

We all depend upon context as a key ingredient for the human skill or attribute we call common sense. When I came here, I passed a million trees. My memory of this experience is not a visual record of tree zero through 999,999. I remembered this part of my trip as, "Gee, lots of trees." Now if I had seen just one of those pine trees upside down, balanced on its tip, roots in the air, I would have remembered it vividly—perhaps even till the day I die. It would be a mathematically trivial change: a 180-degree geometric rotation of one element in a set of a million. But since it violated my previously established and reinforced context of how trees live in forests, it would become instantly memorable to me and anyone else who saw it. The same holds true for a little gray rock, which someone shows me and says "This is the world's most amazing little gray rock." What's so special? "Oh, this particular rock is a part of Mars and it contains what appear to some to be microfossils." All of a sudden I might agree that this is one hell of a little gray rock. Simply speaking, a few words established a completely new context for an otherwise ordinary looking object.

This leads to yet another extraordinary ability of human beings: our ability to deal with abstraction. How is it possible for an animated character to make you cry? When Bambi's mom was murdered by my company, did it make you cry? It was a difficult experience for a lot of people—terrifying for some. How can that be? How can a series of flat, two-dimensional pieces of art make you react that way? How can the flickering phosphors on a television make you think you're seeing something real? Abstraction has been a remarkable genetic gift for human beings. Without it, we would neither be able to read nor write. Unfortunately, it's also why we accept bad computers, bad interfaces and bad interactive experiences.

Simply stated, interoperability is essential for forward progress. What enabled human society to emerge from the first herds of Homo sapiens was human interoperability enabled by the invention and adoption of a common spoken language. Language was the first big protocol-and-standards project. And just as language before it, Internet interoperability protocols are what enabled today's Net—not big, complex messy protocols like those bogging down much of today's desktop software, but simple, basic, low-level, easy-to-implement ones. This was the brilliant gift given to us by the inventors of the Internet, many of whom are among us today.

There is a problem, however. Many people are misinterpreting the great beauty of simple interoperability protocols with blanket approaches to standardized hardware and software. So how did we get ourselves to the stage where many reasonable people think standardization is a reasonable approach to building computer networks? Beats me! Perhaps they're just not paying attention. I think the great accomplishment of the Net is that it allows nonstandardized computers to work effectively together (because of those simple, low-level protocols). This is a very big deal. There wasn't a computer-systems specialist alive on the face of the Earth who 10 years ago would have said that you could network literally millions of PCs, Macs and Unix machines and have them all successfully interoperate, sending and retrieving complex graphics, E-mail, music and multimedia all over the world—not to mention loading and maintaining their own software—and all without an IS person anywhere in sight.

Confusion, and what I believe to be wrongheadedness, also surrounds the art of computer-interface design. At first blush, the concept of a standardized interface seems entirely sensible. Take cars, for example. The fact that we standardized where we place the steering wheel, brake and accelerator was both important and useful. It enabled our world to efficiently learn to drive. The danger is that if we simply take the mantra of standardization and apply it to everything, we run the risk of severely limiting the potential of much of our technology. This gets even worse when we throw in the requirement that these standardized systems must also have intuitive interfaces. I can't think of any high-performance task done by people that uses a standardized or intuitive interface, except of course applications we're all forced to endure on these dumb computers that surround us. There's nothing intuitive about playing a violin, flying a jet, driving a race car, performing a heart transplant or dancing. They all require highly specialized person/machine interfaces that are not at all intuitive to use. A violin isn't intuitive—a kazoo is. Which would you rather hear Itzhak Perlman use to play Mozart? It's okay to say that people have to learn how to effectively use a high-performance interface, just like they have to learn to use language. Even if you love our standardized car-driving interface, try using it to play a violin.

This is not to say that intuitive interfaces are intrinsically evil. They clearly have their uses. For the very simplest, most basic tasks, intuitive is great. But people talk about the intuitive computer interface as if it were the Holy Grail. Natural human interfaces that take advantage of already learned skills (like speech, facial expression and gesture) are great. And it might very well be sensible to use a car interface to drive through cyberspace.

Clearly, natural interfaces like language—and even implants—are going to happen. I believe that it's also inevitable that computers will start understanding the subtleties of what it's like to be a human being—or at least fake it. They're going to follow timing, dramatic pause, intonation, accent, gaze and perspiration—all the things that we sensor-fusing, context-parsing human beings use to help exchange complex thoughts and express our ideas. There will be built-in emotional context overlays so that an appreciation of fear, joy, anger and the like will be constantly factored into how the information is transferred.

Soon we're going to see the next generation of information representations. When I say next generation, I don't mean this nonsense of little dumb icons. Shouldn't an icon that represents something on your desktop that could get you fired look different from something that can't? But now they don't. As human beings, we communicate with more than simple, static, 2D representations—and for good reason. We are naturally brilliant at dealing with—and fusing—complex streams of information. I believe our real challenge lies in getting our systems to provide us with much more and richer information.

In 50 years, computers will evolve into joyful, attractive, aesthetically pleasing objects. They will be something you won't feel you have to cover up when someone comes into your house. They'll be nice to touch. They'll be nice to see. They'll be part of you. Not only will we have the first truly intelligent machines in 50 years, I think we'll be beyond intelligent machines into cooperative communities of machines working in concert to improve our lives (I hope). The important advance from the storytellers' perspective is that we're going to have self-aware machines that will also be aware of you. We will have machines that understand that they are storytellers. We will have storytelling engines embedded in every computer that will be as common as math coprocessors are now.

We hear a lot these days about the inevitable convergence of television and computers. In fact, many say the Internet will replace television. Frankly, I don't think this is an interesting question. I don't think there is any chance that interactive gaming or cybersurfing is going to replace the linear storytelling role that television serves in our present lives. Will storytelling bits be delivered over packet-switched digital networks? Sure, but who cares? What will matter is what has always mattered. Do we care about the story and do those bits represent information that matters to *me?* There seems to be a sense around the Web that television is dead because it isn't interactive. I don't buy it. Linear storytelling, whether in the form of television, radio, theater, film or literature, has survived the test of time because it is a compelling way to express ideas that move us. While I think that eventually we will figure out truly exciting and unique interactive experiences, I believe they will take their place alongside these other forms—not replace them.

None of this will become truly important to us without trust. It's not about trusting your computer, because it's easy to learn to trust machines (or rather to forget to mistrust them). You trust your automobile. If you knew how most of them were built, or how they behave in a 30 mph crash, you might not trust them as much. When you think about the hazards of driving, you realize that the danger doesn't have to do with the automobile, but with the other people on the road. That's going to be the danger in cyberspace and on the Net. You will put your money, your reputation, your life and the welfare of your children on the Web. This will have enormous implications. While we'll never have ubiquitous trust, I think we will eventually be able to carve out enough trusted areas to make most of us (almost) comfortable. Once we have communities of interoperating, self-aware computers talking to other self-aware computers, it would be nice to believe that this extended community is one we can depend on to help—not hurt—us.

Finally, I believe there's one thing that's most important of all—more important than computers, the Internet, crime, poverty, drugs or disease. It's called education. I hope every one of you is as embarrassed as I am that we have allowed the state of education in America to decline to where it is today. On the one hand, we talk about the promise of computers in the classroom; on the other, there is the practical reality that the most advanced technology available in many inner-city schools is the metal detectors that frisk kids for weapons as they enter the building. This is simply not acceptable today or in the future. Our kids are our future!

I'm not suggesting that we should replace our teachers with computers. Those who believe that this is possible, let alone desirable, simply don't know what teachers do. What I do believe is that computers can provide the equivalent of power-steering for our teachers, allowing one great teacher to have a greater reach and touch more kids with their magic.

The fact is, teaching is all about storytelling. Think back and remember your best teachers. I'll bet they were all great storytellers. By combining the art of storytelling and the science of networked computing, we will have a truly remarkable capability to help people learn. Sure there will be issues of haves and have-nots. There always has and there always will be. But just imagine what an electronic book connected to the Net by satellite and powered by sunlight could mean to education. When coupled with the translation and transliteration technology to come, we'll have a device that can break down the two greatest barriers to both storytelling and teaching: the inability to cost-effectively reach people and the lack of a common language. Sure this will be primitive for the next decade. But within 50 years these devices and others not yet envisioned will open windows to worlds that will capture the imagination of our children's children. Each of us has a critical role to play in this process. We must take personal and corporate responsibility to make this happen. Because of inventions like the Internet and the passion of both our artists and scientists, I believe our future is indeed a bright one. I'll see you in 50 years!

The Digital Battlefield

7

James Burke

William Perry has had an extraordinary, varied career. For three years he was responsible for the fact that we all slept safely at night—we Brits as well as you Yanks. He earned a Ph.D. in Mathematics from Penn State, he spent time in the army, he was director of defense electronics at Sylvania General Telephone and was executive vice president of a San Francisco investment banking firm specializing in high-tech. He's been a trustee for the Carnegie Endowment for International Peace; served on a number of government advisory boards, including the President's Foreign Intelligence Advisory Board and the Technical Review Panel of the U.S. Senate Select Committee on Intelligence; he was a member of the Carnegie Commission on Science, Technology and Government; and was on the Committee on International Security and Arms Control of the National Academy of Science. From 1977 to 1981, he was undersecretary of defense for research and engineering, responsible for weapons procurement and for advising the defense secretary on technology, communications, intelligence and atomic energy. After that, he became chairman of Technology Strategies Alliances, a professor in the School of Engineering at Stanford, and co-director of Stanford's Center for International Security and Arms Control.* To top it all, in 1994 the U.S. Senate unanimously voted him secretary of defense. Here to speak about the long-term impact of information technology on international security, it gives me very great pleasure to welcome the Honorable William Perry.

William Perry

Before I look forward, let me first look back. Fifty years ago, when I was an undergraduate at Stanford, the university's electrical engineering program consisted primarily of electrical power. There were few courses in electronics, none in solid-state electronics, none in computer science. So much for anticipating the future.

When I left Stanford, my first job was on a project designing and building a special-purpose computer for the air force. This was going to be used for military navigation through a new scheme called map matching. We used a Gauss–Seidel algorithm to solve 1,000 linear equations and 1,000 unknowns. That was a pretty big task in those days; today it's an easy task for a low-end desktop computer. When our computer was finished, it filled a very large room and consisted of thousands and thousands of vacuum tubes. The good news about those tubes was that on chilly winter days they kept the room warm; the bad news was we couldn't keep them all operating for more than two or three minutes at a stretch, so we had a hard time getting all the way through one whole iteration in our process.

What a difference 50 years makes. During this last half century, we have seen truly amazing changes in technology—especially computer technology. I'm going to talk about the equally amazing changes in national security during that time, the role technology played in making them happen and the interaction between technology and national security in the future.

I went back 50 years in technology, let me go back 50 years in geopolitics. Fifty years ago, Europe and Asia were trying to recover from the devastation effected by World War II—nearly all nations had a shattered economy and an infrastructure in shambles. The Soviet Union alone among those European nations maintained a huge standing army during its recovery period. The United

*Now Stanford's Center for International Security and Cooperation.

States was in a unique position in the world. Having suffered no damage to our infrastructure, and unlike the Soviet Union, we decided to rapidly demobilize our military and made it our priority to build a booming postwar economy. At this critical time, the United States took an action truly unprecedented in history. It offered major assistance to the other nations—friends and foes alike—to help them rebuild their economies. Indeed, it was just 50 years ago this June that George Marshall made his famous speech at Harvard, proposing what came to be known as the Marshall Plan. All the western European nations, including Germany and Italy, gratefully accepted this plan, and the miraculous economic recovery of western Europe was underway—a recovery from which the United States benefited indirectly, but very significantly. But Joseph Stalin turned it down, not only for the Soviet Union, but also for the eastern and central European countries under his control. So began the cold war, with an iron curtain drawn between eastern and western Europe for more than 40 years.

The cold war was one of the most dangerous periods in history. The Soviet Union continued to maintain a large and powerful army, they coerced the nations of eastern Europe into joining them in an alliance called the Warsaw Pact and they invested enormous resources to try to equal or exceed America's capability in nuclear weapons. This military policy was coupled with an aggressive foreign policy designed to expand their influence throughout Europe and indeed throughout the world. For the first few decades of this period, the Soviet Union enjoyed some political success—but at the expense of impoverishing their people and crippling their economy, thereby unwittingly planting the seeds for their eventual collapse. The United States, for its part, was determined to ensure that Soviet expansionism would be unsuccessful and adopted a policy that came to be known as containment. An important part of the containment policy was the formation of military alliances, most notably NATO. At the same time, we were determined not to cripple our own economy by maintaining a huge standing army, so we adopted another policy called deterrence by which we built up an enormously powerful nuclear arsenal that would devastate any nation that attacked us or our allies. But the Soviet Union challenged our interests with proxy wars, which required a response—not with nuclear forces, but with conventional military forces. They had an overwhelming quantitative superiority, so we adopted what came to be called the offset strategy—that is, we decided to maintain a qualitative advantage in our military forces to offset the Soviet advantage in numbers. The key to this strategy was the application of American technological superiority to our weapons systems, and of particular importance was the application of the remarkable developments in computer technology.

Thus, the partnership between American computer technologists and our military, which began during World War II, carried into the cold war. Each of the military services, along with a newly created organization called the Advanced Research Projects Agency (ARPA) developed technical teams that worked with industries and universities to apply American technical know-how to military weapons. This lead to military sponsorship of all the most advanced and innovative programs then underway in computers, space communications, advanced networks, semiconductors and material sciences. Indeed, it is not an exaggeration to say that American leadership in computers, communications and semiconductors in the '50s and '60s stems from the initial military support in R & D in these fields. The results were spectacular, both technically and militarily. During this period, we developed supercomputers used for decryption in the design of nuclear weapons and the simulation of advanced missiles and aircraft. We developed military satellites for global communications, reconnaissance and navigation. We developed ARPANET, the predecessor of the Internet. We flew one of the first integrated circuits on the minuteman missile, and we embedded many computers into weapons systems to give them operational flexibility and precision that were previously unimaginable.

The Soviet Union, in spite of an extensive and expensive effort, was simply unable to compete, and their military fell far behind in technical capability. These three strategies—containment, deterrence and the offset strategy—were the components of a broad holding strategy. I call it a holding strategy because it did not change the geopolitical conditions which led to the cold war. But it did deter another world war, and it did stem Soviet expansionism in the world until internal contradictions in the Soviet system finally caused the Soviet Union to collapse. So the holding strategy worked.

Today, as a result, we are facing an entirely different situation. The Soviet Union has collapsed, the Warsaw Pact is dissolved. Russia and the other successor states to the Soviet Union are trying to form democratic governments with free-market economies. Europe is no longer threatened with a Soviet blitzkrieg. The world is no longer threatened with nuclear holocaust. But it is still a dangerous place. The peaceful transition to market societies is difficult and accompanied by turmoil. As the Italian philosopher Gramsci once said, "The old has died, but the new has not yet been born, and in the meantime a great variety of morbid symptoms appear." These morbid symptoms in eastern Europe caused problems for them and their immediate neighbors. But the morbid symptoms in Russia present a potential danger for the whole world; where there is turmoil, there are still more than 20,000 nuclear weapons. Moreover, rogue nations—Iraq, Iran, North Korea—are trying to obtain nuclear weapons, which would greatly add to the danger they already pose to their neighbors with their large, conventionally armed military forces. So we live in a new era, with new and entirely different security problems than those we faced during the cold war.

What strategies do we have for dealing with these new problems, and how do these strategies affect the computer world? Today, instead of containment of the Warsaw Pact and the Soviet Union, we seek engagement. An example of this policy is the Partnership for Peace in which all 16 NATO nations work with more than 20 eastern and central European nations to demonstrate how the military can support democratic institutions. Another example is the effort to bring about peace in Bosnia, where NATO works cooperatively with a dozen eastern European nations, including Russia. Ten years ago, or even five years ago, who would have believed that in Bosnia today there would be a Russian brigade operating under an American division with an American division commander?

Similarly, instead of practicing deterrence, we practice what I call preventive defense. Preventive defense is like preventive medicine—actions which create the conditions for peace, thereby preventing the need for military conflict. An example of this is what the Defense Department calls the Cooperative Threat Reduction Program, through which we have spent more than a billion dollars of defense funds to help the nuclear states of the former Soviet Union dismantle their cold war and nuclear legacies. In the last four years, this program has led to the dismantlement of 4,000 nuclear weapons, the destruction of 800 launchers and the total elimination of nuclear weapons from three nations—Ukraine, Belarus and Kazakhstan. All of these weapons had been aimed at targets in the United States.

While we have changed our policies of containment and deterrence, we remain resolute in our policy of maintaining military technological superiority, particularly in the computer field. During the cold war, we maintained technological superiority as a way of offsetting the numerical superiority of Soviet ground forces. Fortunately, we never had to put this strategy to a test. But the high-tech weapons systems developed for that purpose were put to a test during Desert Storm. There, we were fighting against a foe equipped with Soviet weapons, but in about equal numbers to the allied forces. U.S. forces were equipped with the new weapons systems developed during the '70s. Known as precision-guided munitions, which used information technology to locate enemy targets on the battlefield—what we called battlefield awareness. They used embedded computers to guide weapons precisely to those targets and used stealth technology to evade enemy weapons. As a consequence, the allied forces won quickly, decisively and with remarkably few casualties. Having seen the results of military dominance in Desert Storm, we liked it, and we decided to keep it. Today, our military strategy calls for maintaining military dominance over any regional power that poses a threat. We do that through our leadership in technology—especially computer technology. This is the same strategy we had during the cold war, but now for a different reason.

Not only has the reason changed, so too has the mode in which we apply this technological superiority. During the '50s, '60s, and '70s, the Defense Department was the principal supporter of R & D for the computer, communications and semiconductor industries. Some of the most significant advances were developed first for military systems—supercomputers, geosynchronous satellites, internets and integrated circuits. In effect, the nation's technology industries were riding on the shoulders of the Defense Department. Today, that has all changed. The technology explosion in computers, communications and semiconductors has led to an amazing new set of products for industry, businesses, schools and the home. Indeed, all of these users are being tied together today by the Web in a way few could have predicted a decade ago. Commercial applications of computers are leading military applications in all of these fields. In fact, computer company revenues dwarf defense revenues. So today, defense is obliged to ride on the shoulders of commercial industry.

This has had a profound effect on the way the Defense Department does business with industry and the way it supports R & D. First, and most importantly, the Department of Defense can no longer support a unique defense industry isolated from the commercial industry. During the '80s, for example, the Defense Department spent years, and more than a billion dollars, developing a family of unique computers. By the time the military development cycle was completed, which took more than 10 years, these computers were obsolete and incompatible with the standards evolving in the computer industry. This example made it clear that our nation should have a single computer industry, which supplies defense and commercial needs alike. That requires the Defense Department to give up its unique specifications and conform to industry standards, to give up its unique buying practices and instead employ the best commercial buying practices.

Transforming our acquisition process is the linchpin of this strategy. It is required if we are to maintain the dominance of our military forces into the next century. That is why acquisition reform was a primary goal of mine when I went to the Pentagon in 1993. I had in mind something that Victor Hugo once wrote: "More powerful than the tread of mighty armies is an idea whose time has come." In 1993, I told our acquisition team that after decades of false starts, acquisition reform was an idea whose time had come, that we really were going to transform the ways we bought goods and services and that they should either get on the team or get out of the way. Most of them enthusiastically got on the team, and we are already seeing the results. For example, we have a pilot program in acquisition reform within the joint direct-attack munitions program, which essentially turns dumb bombs into smart bombs. It does this by adding a global positioning satellite receiver and a computer-based control system to guide the fins on the bomb. Under the old defense acquisition system, which used defense-unique specifications, production requirements and contract rules, the conversion kits would have cost about $42,000 a bomb. Using the new acquisition system, which employs commercial standards, components and buying processors, a conversion kit is costing us about $14,000 a bomb. That's one-third of the original cost, and we are buying tens of thousands of these kits, saving taxpayers about $3 billion.

But acquisition reform is about more than saving money. Most importantly, it will give the Defense Department better and faster access to the new generation of computers, microchips, communications and software. In the past, our procurement process put up barriers to defense contractors using commercial components. These barriers prevented timely access to the information technology that was developing at a breathtaking pace in the commercial marketplace. Acquisition reform breaks down these barriers and speeds up our access to the information technologies by several generations. Transforming this acquisition process means much more than overhauling the way the Defense Department buys its systems, supplies and services. It means a tectonic shift in the Pentagon's relationship with the commercial sector of the American industrial base; it means a tectonic shift in the way we adopt and adapt technology from the commercial sector for military uses; and it means a tectonic shift in the way we get that technology into the hands of the troops. Our major weapons systems take 10 to 15 years to develop, then remain in the inventory for 20 to 40 years. But the computer technology that most influences their competitive advantage changes every two or three years. So we need to evolve a strategy that keeps major weapons systems in the field for several decades, but updates them every few years with new information technology that can provide real military advantage. At Fort Hood, a large-scale experimental program is underway to do just that. The army has chosen Fort Hood and the 4th Infantry Division to experiment with ways to develop Force 21, the computerized battlefield. I witnessed two of those experiments, and both hold promise to dramatically improve the way we adopt and adapt computer technology for military uses. The concept behind the first experiment was simple: Insert digital subsystems into our current weapons systems—tanks, artillery, helicopters—thereby giving them a quantum increase in capability. An aggregate taken from all these platforms forms a system of systems—an integrated network of powerful computers and high-speed communications. This system of systems will transform the way commanders and troops see and communicate on the battlefield. In the past, information was passed around by radio conversations or typewritten messages. The result was that the commander received only a fraction of the information he could really use in combat—and he got it a day late. With the system of systems, commanders have the ability to send and receive digital bursts of critical information about the location of all enemy and friendly forces; the rate of usage of food, fuel and ammo; and the progress of current operations and the planning for future ones.

The effect on combat operations will be revolutionary. Every commander will have battlefield awareness—a constant, complete, 3D picture of the battlefield. Every soldier will have the information needed to carry out the commander's orders, and an entire army division will move as one integrated battle system. How does this work in practice? When a tank commander spots enemy forces, he will have a choice: He can engage the enemy with weapons on his tanks, or he can call in nearby attack helicopters, artillery, strike aircraft and naval gunfire. Because of digital technology and the constant flow of battle information to all combatants, these other units will see exactly what the tank commander sees. Any one of them, or any combination of them, will be able to respond with equal precision. As the combat is underway, the supporting logistics unit will be monitoring the ammo usage. It will conduct resupply at the time and amount needed, thereby reducing the huge logistics tail needed to support combat operations. This system of systems is a brilliant application of information technology to achieve battlefield dominance without designing all-new weapons platforms. When I visited Fort Hood, I saw the future—and it works.

In addition to incorporating new information technology into existing weapon systems, we must develop the military tactics to get the most out of the technology in a combat environment, and develop the training so our soldiers can maximize the use of advanced technology. Is it possible to develop the three Ts of forced dominance—technology, tactics and training—concurrently? Trying to achieve this led to a second experiment at Fort Hood. Until now, we developed these three Ts one after the other. This created long periods of acquisition, test and evaluation, fielding and tactics development. Then, after decades, we would get around to training the soldiers to use the new systems. But if we fuse the sequential process into an integrated, concurrent process, we could cut years off the time between developing new technology and getting it into the hands of our troops. The army's answer at Fort Hood was to create a process action team comprised of the experts who developed the three Ts—the acquirers from the army material command, the requirers from the training and doctrine command, the users from the forces command and the builders from the defense industry. Against all odds and contrary to all traditions, this team is achieving success. Soldiers prove time and time again that they are critical members of these teams. The critics used to say that our troops would not be smart enough to handle high technology. The critics were wrong. Soldiers not only operate and maintain the latest digital technology, they find new and ingenious ways to employ it. As a result of the involvement of our soldiers and other experts in the Force 21 process, the three Ts are all approaching combat-ready standards at the same time and in a realistic environment. This puts us years ahead of where we would have been had we followed the traditional sequential developmental process. For the first time, the military development cycle was in synch with the development cycle in the computer industry.

But the real test of the Force 21 process and the new system of systems is whether they can succeed in actual combat conditions—in the fog of war. To find out, a brigade from the 4th Infantry Division will take the system of systems to the national training center at Fort Irwin, California, which represents the ultimate in ground combat training, including well-equipped, well-trained mock enemy forces. The training is realistic and tough and aims to make the scrimmage tougher than the game. I saw how advanced computer technology serves the troops during my last visit to Bosnia. As I toured one of our base camps, a young soldier showed me a computer terminal hooked up to a satellite dish. Using this system, the troops download high-resolution digital imagery of the Bosnian countryside. The imagery is routinely collected over Bosnia by our satellites, reconnaissance aircraft and drones. It is then fused and analyzed in England and sent to the computer terminal in Bosnia within hours. This imagery helps our troops immediately see any movement of heavy weapons in violation of the peace agreement as well as the gathering of troops or mobs that could harass them on patrol. This makes the troops more effective in their missions and saves lives. It demonstrates why transforming our acquisition process is so important to me, and why I am so proud of how far we have transformed this process during the last few years.

In sum, as we enter the 21st century, our security strategy must change to reflect the dramatic geopolitical changes of the last decade and the dramatic technological changes of the last few decades. We will proceed with a policy of engagement with our former cold war adversaries; a highlight of this strategy will be the Partnership for Peace Program. We will proceed with a policy of preventive defense to create the conditions for peace, especially by assisting nations that are trying to establish new, democratic societies; a highlight of this strategy will be the Cooperative Threat Reduction Program for reducing the nuclear legacy of the cold war.

Finally, we will proceed with a policy of maintaining U.S. military dominance with the timely application of our computer technology. To do this in the face of the remarkable technological advances in the computer industry today requires a complete transformation of the way the Pentagon does business. That transformation has already met with significant success. Winston Churchill once said, "You can always count on Americans to do the right thing, after having first exhausted all other alternatives." We have indeed made many false starts in the last few decades trying to transform the way we manage our defense business. But this time we are doing the right thing—and computer technology is leading the way.

FERNANDO FLORES

Entering the Age of Convenience

8

James Burke

Our next speaker is regarded by many as one of the leading scholars in the field of management today. He was a provost at the University of Santiago, Chile, at the tender age of 27. He was also technical manager of a large conglomerate, director of the Inter-American Bank, president of the Chilean Institute of Technology and director of a project to apply cybernetics to management. In 1976, he came to Palo Alto, California, where he did research at Stanford, and where he met and co-authored the classic book, *Understanding Computers and Cognition* with Terry Winograd. In 1979, he received his Ph.D. from Berkeley for a dissertation called "Management and Communication in the Office of the Future." This speaker has devoted more than 30 years to the study of how human beings work together. His field combines the disciplines of linguistics, philosophy, computer science and management as they are applied to the vital matter of doing deals. At Business Design Associates, which he founded, his multidisciplinary approach to business brings a fascinating new view of what it is we all do and how we do it. Here to speak on the long-term impact of information technologies on business, please welcome Fernando Flores.

Fernando Flores

I have been asked to talk about the future of computing and business in the next 50 years. I believe all computers are communication, and all business is communication. The question is how you talk about that. In my opinion, computer scientists and technicians talk about communication in a language that businesspeople don't speak, and businesspeople talk in a language that the others don't speak. That's a big gap that has enormous consequences. My contribution here is to try to bridge this gap.

We are coming to understand what people understood in the Middle Ages. We live in the world of uncertainty. That's the human world. Control is very limited. We also begin to understand that more freedom means more uncertainty. We fight for freedom and thus we are increasing uncertainty. In a certain way, this adds more innovation—and more innovation, by definition, brings more unpredictability. We are entering a world of more uncertainty, and we need to develop emotions to live with that.

What I'm going to talk about is what gives us the right to do this kind of speculation. First, in certain circumstances being a storyteller allows you to talk about how marginal practices go to the center. Marginal practice, in my opinion, is a way to look at the future already in the past—that is our present. Second, you can articulate the future's potential by arguing that fundamental occurrences have been discovered but not yet exploited. You see this in science all the time. DNA was a hypothesis of fundamental occurrence; four basic sequences explain everything. The real understanding—not only intellectual but pragmatic understanding—produced in 1975 what we now call biotechnology. So biotechnology was impossible until scientists grasped the structure of DNA. I believe we will not see this technology make a positive contribution to humankind until we further investigate what the human mind is.

I want to tell you a little about my prejudices. I was educated as an engineer and a philosopher. I was trained in the United States in English and German philosophy, and I also live in Chile. I was a political prisoner for a while, and I was minister of finance at 28 years old. What I learned was basically two things: how trust can be broken, and how patience can be gained. The other important experience in my life has been being a father and being married for 35 years and having six children and five grandchildren. This does something to you. I will talk from that prejudice, because I have no other way to talk. Finally to avoid having my prejudice kill me, I have friends who fight with me, discuss with me and give me certain direction.

What is going on in cyberspace is not a technological revolution. It's a change in the way human beings relate. Human beings have essential relationships, and when we change those relationships we change the way we work and the way we do business. Business is a special social relationship. We need to investigate how social relationships are changing in business, and how technology is affecting them and vice versa. But this extends further than business. We are also changing the way we relate to ourselves. It's a big transformation of the self, and that's the central issue. Some of the speakers have asked this question: "What kind of human being will we begin to be?" That's how big the question is—so big that I will deal with only part of it.

This industry has a tendency to treat communication and information as if they're the same thing. I see this again and again, but they're two separate phenomena. Information has to do with what is present and can be asserted. If you have recorded you can assert. But communication has to do with living together successfully. Good communication has to do with the intentional coordination of actions. I believe that Web technology is shifting the space of human communication. It is producing communities that could not be formed before and transactions that could not happen before. That's what's really interesting. There are new political organizations that are going to derive from all this.

In doing this presentation, I have divided historical time into three ages, and I'm going to describe two now. I call one the age of need and the other the age of convenience. Within the age of need there is a technological component, which I will call the age of information. In the age of convenience I will call the technological component the coordination of commitment. Let me explain what I mean. Need is a special kind of relationship that businesses and people have. People have desires; companies satisfy those desires by making products and use planning to improve them. You still see that relationship in certain kinds of products, but less every day. I can't go into a supermarket here and find bread. People ask, "What kind of bread are you looking for? What kind of health do you want to cultivate?" In the age of need we were still defined by our desires, and what people needed were transactions. Information was the recording of those transactions, but the deal was happening in the conversation; the transaction happened afterward. The transaction was important for control and for the IRS—it was not important for people. That produced massive complexity, and that's why when the computer arrived, at first it was a data-processing machine. It was years before the computer was used for military purposes. It was a marginal invention brought to business by businesspeople. Then it became a super-tabulating machine. To this day we adopt this terminology—that everything is a transaction, everything is information.

But let us also look at certain dreams from that age. When I was at the university, we had the following equation: more data, more knowledge, more power. We have discovered that this is not necessarily true. Today we are living in the age of inflation of data. We need less data. We are talking about filters; we are talking about agents. We are beginning to distinguish between knowledge and power. They are interconnected, but power has to do with access. We also hope that more powerful, rational models with the right data are going to be better for us. I was educated in operations research. Some of my friends were educated in artificial intelligence. We learned to produce models, we learned to produce certain expert systems. But the big problem of human rationality today looks totally different. It looks like a dream. What we have yet to discover—the essence of human rationality, emotion and intelligence—has to do with human interaction. The moment the network appeared, human interaction became the center. In computers today, a new kind of worker is emerging—not a knowledge worker but an interactive worker.

As for information technology, I can speculate that in the next 10 to 25 years we will enter the age of convenience, which consists of the following: People don't just want products; people want products that come with service, that come with training, that come with maintenance, that come on time and that appear when we need them. Timing and location become very important. For this to happen now, people need companies to promise unconditional satisfaction. People don't need objects; objects are trivial—they can be delivered to someone else. Domino's Pizza doesn't produce better cheese or ham; it produces a promise to deliver within 30 minutes. Federal Express doesn't produce quick delivery; it produces a promise of delivery.

We are beginning to see this all around. I am working with banks, engineering companies and a manufacturer, all of which have the same problem: how to organize, coordinate and use this technology. My opinion is that the explosion of business reengineering was the catalyst. The companies began to reduce the flow of paper and are now beginning to see that the central issue is people. We cannot do that transformation without bringing ethics to the center, because a commitment is a flexible bond in time. The essence of commitment is that it happens, it has a future and during this time we are bonded. But it's not a rigid commitment; it's a flexible one. People are expecting something that can be attended, like an airline flight, or can be canceled without cost to us. Companies have to be ready for that. I see the world going in that direction. People are not willing to tolerate delays or rules that are unnecessary nor will they put up with immense bureaucratic inventories. The convenience for customers is also convenient for capital. People don't want working capital; working capital equals zero. They want convenient coordination. It's better for the client, better for service, better for people.

Convenient coordination is going to require two things that have already happened: First, we need systems that support the coordination of commitment. It's a theory of work. We don't need electronic mail, we need tools for work. Work involves dealing responsibly from a network of commitment that defines us. Second, we need the ethical transformation of people. One of my big pleasures is that I look Mexican, but I am Chilean. I am also European, yet American. Every time I go to work for a big company, particularly in Germany or Switzerland that have reputations for punctuality and rules, they bring me—a Mexican. You know what they bring me for? Impeccability. Isn't that funny?

Why is this happening? Because the old, rigid rule doesn't work anymore. We need flexible impeccability. That means good listening and no listing requirements. Every company I see listing requirements has failed. It is very clear that Microsoft didn't invent *Windows* by listing requirements. It invented a hit. Intel invented a product that's a hit. Today, the leading companies are leading the space of reality for people. The Germans are not doing that; the Japanese are not doing that—and they're having problems in the '90s. This is the age of convenience at the customer level; it requires exquisite coordination of commitment at the level of interaction.

I want to say something for my fellow Latin Americans here. We don't have a chance in the future if we don't bring our culture to produce. We will still be selling cheap labor, and that will perpetuate poverty forever. For America, that's going to be a big problem for defense. For the goodness of this country, and for the goodness of us, we need to participate in this culture.

The age of convenience, the age of coordination and the age of flexibility are going to present some problems, which are already beginning to crop up. One of them is a lack of direction. People don't know what to do next. People in the age of need used to look to the past. That's the way people did strategic planning. What was done next was supposed to be more and better—and different. That's all. But now they're seeing a radical difference. I like the spirit of Intel Chairman Andrew Grove's book, *Only the Paranoid Can Survive,* but I don't know about the word "paranoid." I have seen what Microsoft has created in response to the Net; that's what needs to be done.

This is a moment of tectonic change of the whole ecological niche in which your customers, providers, suppliers and competitors change as well. To do this you need vision, but vision is not strategic planning; vision is not expertise. I just went to see a company in this country, which held a conference on innovation, and there were a lot of experts. One of them asked, "Why don't you run your company like an architectural company?" Another said, "What you need to do is position your brand." You know, there are still companies that talk about brand. It looked interesting. Another person said, "You are really good at moving materials. What you need to do is operate as a logistics company." I asked, "How do you feel?" They said, "We feel more disoriented." Right. That's what lack of direction is. We cannot find our future by listening to experts. We need experts to open horizons, but for inventing our future, we need to have a strong identity that is flexible.

This is where we enter the third age: the age of shifting concern. What is the difference between a concern and a need? A concern is when you listen to the long-term interests of people, which include, for example, health, education, having a job and raising a family. But concern is not a product; it is not a need. Concerns need to be interpreted and packaged and the package is changing all the time. What are banks today? They used to be a very important, clear, well-defined business. Today a bank is a bunch of computers connected by terminals. Is it a software company? No. But there's another part of the bank that used to take care of relationships. Those are now being handled by the computers. The people running the banks today are all thinking, "What are we going to be? How are we going to do it?" That moves into the issue of identities. We need to deal with that issue as individuals and as persons.

My whole claim is that we cannot invent our identity by looking at what we did in the past. We used to do that in the age of need. Today, we need to take a stand about the future. The central thing is to discover who we are. Every good entrepreneur and innovator I know is a material historian, because without looking at history we don't understand our nature. A great many of our practices—the ones we do in the background, the ones we don't think about—come from the past. The other day my colleague Alan Kay, a computer scientist and historian, asked me, "Do you know why books are numbered?" It had to do with Erasmus. When people began to have discussions, they wanted to refer to certain pages. At that moment in history, books needed to be numbered. So, we need to discover our relationship to the past, where those relationships are going, where the industry's going, where other industries are going and we need to fall in love with what I call "marginal practice."

Marginal practice is something that someone is doing in another key area, or something that goes on in another industry. To be leaders we must fall in love with the cultivation of marginal practice and accept that we cannot be good at everything. That's why we are afraid. Without direction we are scattered. We call it information overload, which is a consequence of a lack of clarity of who we are. This, then, is a skill we need to develop in school and other places. What is the first thing we need to do to develop a strong identity? Develop it with others. When I look at Silicon Valley, I admire the kind of culture that has evolved there, how people can have companies that compete and cooperate and define each other. The biggest leader there started small, and built relationships.

We need to develop strong relationships, like marriages, for the future. In work, marriage is not a lifetime contract despite what is sometimes said in the pronouncement; it is a serious, open engagement that doesn't define conditions. It's not about need or convenience. It's about being partners in good and bad times for the length of the partnership. This is going to be an important principle that will be extended to the rest of the economy. For example, in fashion, Laura Ashley and Ralph Lauren have loyal clients. They are married with style.

We will need to find our niche in what I call an ecological web of clients, an ecological web of allies. We cannot do these things without finding a role for ourselves, and the first thing we must do is project our identities. Like Bran Ferren, I believe the central skill human beings need if they are to become leaders, entrepreneurs and good persons is the ability to be storytellers—storytellers who bring care and concern to others. We need to be seductive authors for the future. If we become this kind of being, we will be able to deal with the problems of directionlessness and scattering. If we don't, one possible result is what I call the hyperflexible, postmodern society; we are going to be very weird people connected to computers—never connected with people. We are going to lose social skills, and we are going to lose affection and other interesting things. I don't like this future. I want to fight for another one.

I don't see another industry that has more relevance to what's happening today. We are designing cities, we're designing economies, we're designing forms of life. My plea to you is to take seriously the notion that human beings are like narratives. We are what we are because of what people say about us and because we say and do certain things. What are the two aspects of humanity that allow me to be sure of what I'm saying? Whatever human beings are, they are human beings making commitments—offering promises, requests, declines, counter-offers and marriages. If your boss sends an electronic mail saying, "You're fired," that's not information; it's a lot more important. If you make an offer of marriage to your future wife or husband, that's not information. Language will also make a difference in people's lives. With language we invent and discover worlds with other people. What kind of storytellers are we? We are storytellers of world openings and commitment and, yes, we are also storytellers of information.

The last thing I want to say to you is that I belong to Chile, a country that is part of the economic success of the last 10 years, but which still belongs to the third world. There are a lot of people who are going to be affected by this new technology, but these people have not yet arrived at the age of need. They are still in the age of power, misery, injustice and lack of dignity. We have the responsibility to think about that, but I suppose that's an issue for another conference.

VINTON G. CERF

In the Belly of the Net

9

James Burke

Our next speaker is one of those rare people who was wired before anybody knew what that word meant. That's not surprising, because if you do any surfing at all these days, it is thanks mostly to him. He's known throughout the industry as the Father of the Internet, although he's far too modest to acknowledge the paternity. Anyway, he earned a Ph.D. in Computer Science at U.C.L.A. From 1976 to 1982, during his time with the Defense Department's Advanced Research Projects Agency (ARPA), he played a major role in sponsoring the development of Internet-related data-packet technologies and co-developed Transmission Control Protocol/Internet Protocol (TCP/IP), the working protocol that links all the diverse university, government and commercial data networks. That lets us spend hours of fun and frustration not finding what we don't know is there. But that's not his fault. Anyway, from 1982 to 1986, he was vice-president of MCI Digital Information Services, serving as the chief engineer of MCI Mail. Recently, he was vice president of the Corporation for National Research Initiatives, where he conducted national research efforts on information infrastructure technologies. Today, he's senior vice president of Internet architecture at MCI Communications,* and I can tell you he's a jolly nice chap. His task in talking about the next 50 years, however, is a little tougher than most of the speakers because 50 years on the Web is like centuries anywhere else. So here to speculate on the world of the future where the Net is everywhere and everybody, please welcome Vint Cerf.

Vinton G. Cerf

I wish I was eight years old, because I would really like to see the Internet in 60 years. Kids today are going to see an Internet that we can't even imagine—an endless frontier. Well, you're right James. It is a terrible challenge to try to predict anything 50 years into the future. But you know, fools rush in where angels fear to tread, so here I am. I really do wish I was eight years old, because it would give me a chance to see what's going to happen over the next 40 or 50 years, but since I'm not, I'm going to take the opportunity to speculate about it anyway. You know, we've all had visions of what the future will be like. I think Arthur Clarke was right on the mark when he envisioned the possibility of space communications—and that was around 1947. On the other hand, HAL didn't quite appear when everybody expected him to, so sometimes we're not very accurate in our predictions about the future.

In some ways, it seems like nothing happens overnight and takes an incredibly long time to mature. On the other hand, sometimes it seems like everything happens overnight. Fax, for example, was invented in the mid-1800s, but it didn't really come to fruition until the mid-1900s when it turned out that in Japan, the Kanji character set was not very conducive to keyboarding. For fast communication, the only technology that came to mind was facsimile; and once there was a ubiquitous telephone system on which to base fax machines and the standards that allowed them to work, it was entirely possible to overcome this keyboarding problem. Things have gotten better, and it is now possible to keyboard Kanji in ways that had not been possible before. It's quite fascinating. You type the phonetics for what you want, and a possible Kanji character pops up. Then you can cycle through whichever one might be appropriate. It's a little awkward, but it's better than not having that ability at all.

*Now MCI WorldCom

Television happened the same way. It was designed in the 1920s, and yet we didn't see it emerge as an important medium until the 1950s.

For some of us, the Internet has been a long time coming. It was 1973 when Bob Kahn and I first did the original design work while I was at Stanford University. Bob was at ARPA, and many people were working on this system—thousands of them since. Yet it has taken more than 20 years to see this system come to fruition. One sometimes imagines, though, that others in the world don't see it that way. I think it's fair to say that the public didn't notice that the Internet was anything of interest until about 1994, when the World Wide Web became so visible thanks to Netscape Communications and Marc Andreessen. Bill Gates discovered it in 1995, and to give Bill credit, he took his company and aimed it squarely at Internet applications.

The other thing I think is very important to all of us is that the Internet is not driven from the inside. I supply a piece of the Internet at MCI, or my company supplies it, but in fact it's being driven by forces that are outside, by people who find applications and just want to try things out to see if they work. I'm mortally terrified of the graduate student who doesn't give a damn about profit and just wants to see whether he can make a new application function. That's where things like Internet telephony come from and videoconferencing such as CU-See-Me and the like. In fact, it is from that wellspring of "just try it" that much of what happens in the Internet comes.

Just to give you some sense of how far technology has come, I want to tell you a personal story about my wife, Sigrid, who lost her hearing when she was three years old. That was about 50 years ago, and I know she'll forgive me for telling you that. For those 50 years she was profoundly deaf. She could lip-read and spoke almost normally, but last year, after doing some research for several months on the Net, she discovered a technology called cochlear implants. This is something that literally replaces the inner ear. It goes into and electronically interfaces with the auditory nerve. It takes sound from a microphone, or any other sound source, and goes into a speech processor, which she wears stuffed in her bra sometimes—the only 100 megahertz boob in the country. Then she has a wire from the speech processor to an inductive coupler, which interconnects with this electronic inner ear. If someone had come into my office and said, "This is what we're going to do," I would have thrown them out of the building and said, "Come back when you have something sensible." But this thing actually works. It works so well that she uses the telephone, she watches television, she turns off the captions—much to my dismay, because I'm still hearing impaired. Sigrid is now hearing repaired, as nearly as we can make out. She makes use of this thing in theaters, where they have infrared pickups for people with hearing impairments, except she detects the infrared on the little gadget hanging around her neck and then plugs it into the speech processor. When she's on an airplane, she uses a patch cable to plug the speech processor into the armrest—and she doesn't need one of those earphones that they come around peddling for $4. This is truly amazing. Before all of this, she spent eight months tracking other people to see how they did with these cochlear implants. After she concluded that this looked like a good thing, she went up to Johns Hopkins University Hospital. They do this on an outpatient basis, so she went in the morning and came back that afternoon with the implant. It wasn't activated yet because they have to wait for it to heal. Four weeks later, she went back and they activated her. I get very teary when I think about this, but about 20 minutes after they turned this thing on, she called me and we talked to each other on the phone for the first time in 35 years.

Wow, so technology works. We're starting to see the telephony world and the Internet world as part of a common fabric. At MCI, we announced a new technology called VAULT. What we're trying to do is take the canvases on which we've been painting telephony applications and services for 125 years, and canvases on which we've been painting Internet technology, products and services for maybe 10 years and tear those canvases apart thread by thread. Then we want to reweave them together into one common canvas on which we'll now paint products and services that could not have been built except for the commingling of the two. I can see network-independent services that are no longer sensitive to what kind of transport medium they are going over. The Internet protocol was designed to go over any transmission medium, which is why I have a tee-shirt that reads, "IP on everything." I even got one for my dog to wear when we take her for a walk.

So, what we are really seeing in the late stages of the 20th century, is a new birth of telecommunications. I often wonder what it would be like if this was 1897 and the telephone system was about the same age as the Internet. Kitty Hawk is six years into the future. Einstein won't write his paper for another two years after that. What Kitty Hawks and Einsteinian papers await us in just a few years time? I don't know the answer to that. One thing I can tell you, though, is that the term "Internet telephony" is about as accurate as "horseless carriage." We all understand that it is backward-looking instead of forward-looking. What's really going to drive the evolution of this system is interoperability. That's fundamental. Customers want interoperability. To get it, they want open standards. New application possibilities open up when there are things in common. Metcalfe's law, which holds that the network's value increases as a function of the square of the number of participants, is at work here. So it's fundamental that even if a company wants to go off in a direction which makes things incompatible—because it's doing product differentiation (an understandable desire)—it must realize that eventually the customer is going to force it back to some commonality so that things work together and the user is not locked in.

Of course, every time we create a new kind of standard, we create a new platform on which new products and services can be built. The World Wide Web has become an enormous platform for new products and service development. The Java language from Sun Microsystems is similarly creating a new kind of platform. Even interactive computer-based games may turn into yet another platform.

We are a society that wants everything, and wants it now. This is the American instant-gratification ethic, and the Internet is beginning to respond to it. People want information, and they want it now. Not only that, once they get the information—if it happens to be about a product or a service—they want the product or the service now, thank you very much. That's why you get companies like amazon.com, which has no physical inventory as I understand it. You place an order for any book in print, and they make the arrangements and get it to you the next day or the day after. I suspect that as we start creating products that can be delivered electronically, this instant gratification will become even more pronounced. Fortunately, there are many things that can't be delivered that way unless we get to the beam-me-up-Scotty stage, and I don't predict that within the next 50 years. So there will still be a great deal of need for people to get out and do physical things. We're not going to have a population trapped behind its display panels forever and ever. There's too much desire for human interaction for that.

We're certainly going to see, however, a lot more telecommuting than we have in the past, because as the network becomes more pervasive, has a higher bandwidth and better functionality, it will become more and more possible to be effective at home and on the road. But there's a side effect to this: As companies introduce this technology into their internal operations, and people begin to use it on a regular basis, they may find opportunities to play while working. For some companies that's an issue—even for some universities. *Doom* and *Quake* are popular games, but they soak up a lot of bandwidth in campusland, and some universities have had to say, "Wait a minute. How about not doing this during the day?" Perhaps we will see micropayments showing up as a real element of our economy. I'm still not 100 percent sure about that, but I do remember the penny novel from a century ago, and I wonder whether we will see its rebirth in the Internet environment.

There is a little cabal that's beginning to form, and the kids who love the Internet are going to hate this. Think about how people deal with the educational system today. There's Johnny's mommy and Johnny's teacher. Johnny comes home and says, "I don't have any homework." As the Internet gets more pervasive, mommy will say, "Yes you do. I looked on your teacher's Web page. You have some history homework tonight, and you have some math to do and I saw two other homework problems." Then when Johnny's teacher and Johnny's mommy have to get together, instead of coordinating a meeting at 7 o'clock in the morning when they're both half asleep, they can send E-mail back and forth. Johnny is going to hate this, and if there are any Johnnys in the crowd, it wasn't me—it was some other guy who came up with the idea.

Something else is going to happen: the infusion of networking into appliances that we don't normally think of as being networked. I am a great proponent of being continuously connected to these systems, but I really hate the idea of having to do a circuit-switched call to get connected to the Internet. What I want is dedicated access, and I want my packets to go whenever they have to go, thank you very much. At some point I hope it happens. One of the things that may help us get there is the introduction of Internet-based computing inside appliances. For example, can you imagine a scale that you step on, lights up and tells you what your weight is and then sends that information to a local medical-record database, to the doctor and makes some snide remark about what your weight was yesterday compared with today? Or alternatively, it sends information about current weight trends to the refrigerator, which has also become very smart and refuses to open because it knows you're on a diet.

Another thing which is almost certain to happen is the advent of smart sensors. We'll probably see this first in medical technology, where a lot of non-invasive methods already have been developed to determine your state of health. There's even an example of an intelligent toilet, invented in Japan, that analyzes your bodily waste and lets you know whether there's anything serious going on there. The other possibility for smart sensors comes from a more prosaic example: Today, it's very common for us to breeze through doors without even thinking about opening them because we expect them to open automatically. I imagine one could invent more elaborate kinds of sensing equipment to go with the furniture that figures out whether it should get warmer or colder or adapt the shape of the cushion or something like that.

One thing I believe is going to happen is that reading glasses will become a substitute for the displays we use today. I justify this on the following grounds: If you've ever thought about trying to have breakfast in the morning while reading the newspaper on your laptop, do you put it in the middle of the scrambled eggs? There isn't any room on the table, so one answer is you put one big flat-panel display on the wall, and then everybody fights over whether to view the funnies, the business section or Dear Abby. No, I think the answer will be reading glasses, very lightweight gadgets that project a virtual image which appears to be within arms length. They might contain a little finger mouse that operates in three dimensions, allowing you to point and click in this image. This device will be very helpful for people who work on airplanes because in the past we had really crappy displays, and someone sitting next to you couldn't read it because it was viewable only within a small angle. Now there are high-quality multiscan displays that allow anyone sitting in the next seat to eavesdrop on what you're doing. The glasses would be a protective measure, so I'm actually looking forward to their invention. I thought that would happen in the 21st century. I'm embarrassed to tell you that my friend Michael Dertouzos sent a book he just wrote called *What Will Be,* saying these glasses have already been developed in his computer-science lab at MIT. So much for my ability to project into the future.

Another fascinating development is the commingling of the various media and the Internet. We're starting to see telephony, video and radio—or their analogs—going through the Net. Now, it's important to understand that the Internet, as it currently exists, does not have the capacity to handle very much of this. However, in the long-term, I don't see any technical reason why we can't scale the system to handle a substantial amount of this kind of traffic. It's going to take some serious work inside the routers or the switches, for example, to distinguish high-priority traffic from low-priority traffic. As for Internet video, it displays a 2 × 2-inch screen that runs about 10 frames a second. Congratulations! You've turned your $5,000 laptop into a 1928 television set! Radio works better over the Internet because it doesn't make quite the same demand on bandwidth, but it does something funny to the radio business. It used to be that the size of a radio-listening audience was a function of how much power could be put out of the antenna. Now, audience size is a function of who's interested and where around the world. That may change the economics of radio as much as it does anything else. It's also possible to take conventional media, like television and radio, and inject them into Internet packets. Intel has created a system called Intercast for doing that. So, we're seeing the Internet interlacing itself with all these different mediums. I think you'll see some mutual reinforcement happening; trucks going down the street emblazoned with "www.UPS.com," Internet advertisements on TV or soap opera summaries on the Net. One is tempted to speculate that paper will become a specialty item, although as I'm sitting in the men's room with my laptop, I'm not sure I'll think the paperless society is the best idea I've heard.

This whole evolution is important, not only because it is a new mode of human communication, but because computers will be able to communicate autonomously on our behalf. I carry an awful lot of stuff with me—pagers, hearing aids, batteries, rechargers, portable telephones—and I'm hoping that a lot of these will get merged into one or a smaller number of devices. We can certainly predict that we will be wearing clothing that does computing and communications. Of course, we'll need a new vocabulary for that. Some people will discover that by having television coming in via the Internet, they may be able to carry out activities other than just being a couch potato. If that happens while advertisements are being sent, we may also see a disconnect between the economics of television as we know it today, and what it will be in the future.

Speech recognition is an area that has experienced a great deal of hope and not as much success. There is, however, progress being made in very narrow areas where only a small vocabulary has to be detected. That is where we should see some serious results—certainly by the year 2047.

Finally, the real progress: Once your VCR becomes a part of the Internet, that's where it will set the time, and you'll get rid of that damned blinking 12. Moreover, for those of us who can't figure out how to program the thing, we'll see a nice Web page come up with a set of television programs on it, click on the ones we want to record and the system will tell the VCR when to do it. End of problem.

There will almost certainly be a lot of multi-party gaming going on, and there's a secret here: Nintendo, Sega and others are already working with Microsoft and others to build Internet-capable video games. This is going to lead to real video conferencing, not the stuff that all the industries—including us—have tried to sell. Think about what goes on in those video games. You're shooting at each other, but wouldn't it be neat if you could hear the other guy go, "Ahh!" when you shoot him out of the sky? There probably are going to be microphones as well as speakers on these things. And then it would be really cool if you could see the other players, too. So you get these cheap little connective television cameras to plug in. Now there are a bunch of people all screaming and shooting at each other. Sounds a lot like a video conference, doesn't it? So this is where video conferencing is really going to come from.

I think we will certainly see experiments with virtual shopping and banking and those will evolve very quickly in ways based on user preferences. We may see the refinement of digital actors. Today's special-effects methods are such that you can't really tell what's real and what isn't anymore. We're going to have to look for some new techniques to distinguish visual fact from visual fiction.

There's a big question about whether the Internet's really going to survive. There are going to be some stumbling blocks, but there are solutions. If you're worried about having enough address space, the 32-bit address space, which I freely admit I screwed up on, should have been bigger. It is going to be 128 bits, which seems big enough for every electron in the universe to have a host on the Internet if it wants to. The routing technology has to change to be able to handle multiple qualities of service. That's coming. There are such things as really big routers—BFRs—that can handle this kind of traffic and which let us build big nets—BFNs.

If you're worried about traffic and fiber capacity, don't. There's lots left. We're already operating a 40-gigabit-per-second fiber in our underlying commercial fiber network. We can already see 100 gigabit per second coming and terabit per second has been demonstrated in the lab. The issue here is that it's tough to produce a router that can keep up with that. If you're worried about costs, competition is going to drive them down as well as prices. Once you get an infrastructure like this, where people's livelihoods—and possibly their lives—are dependent on it, regulation in the public good almost certainly will occur. On the other hand, things like censorship won't work, and I credit John Gilmore of Sun Microsystems with this wonderful quote: "The Internet interprets censorship as damage and routes around it." As well it should.

So we can anticipate several things: One of them is that there will be more devices on the network than people because computers will be in the landscape, houses, all the furniture and everything else. There will be lots of computer-mediated communication. I can tell you that the Internet is going to solve Alzheimer's disease—well, at least my Alzheimer's. A few weeks ago, I couldn't find something on my hard disk and I thought, "You know, this file might have been published." So I went out on the Web, did an AltaVista search, and the third hit was the file I was looking for. We can definitely anticipate a lot of interaction between intelligent agents throughout the network. So, I will make a prediction now and then stand down: I will not live to see what happens in 2047, but I *will* live to regret the timidity of all the predictions I've tried to make.

BRENDA LAUREL

When Computers Become Human

10

James Burke

Our next speaker started out in the personal computer business as a programmer, software designer, marketer, producer and researcher. Her academic background is in theater, with a Ph.D. from Ohio State, so either she's a gamekeeper turned poacher, or a poacher turned gamekeeper. Either way, she's going to make you think. She began the main body of her work when she went to Atari in 1982 and started developing software architectures and interfaces for computer-based interactive fantasy systems. Since 1987, she's worked as a consultant in interactive entertainment and interface designs for clients such as Apple, American Interactive Media, Fujitsu, Carnegie Mellon, Lucas Films, Sony and Paramount. In 1990, she founded Telepresence Research to develop virtual reality and remote-presence technology. In 1992, she joined Interval Research Corporation, where she is now, as a member of the research staff and led a three-year project to explore the influence of age and gender on children's play. In 1996, she co-founded and became vice president for design at a new company called Purple Moon, where she's working on a project to develop interactive systems for young girls based on her earlier research. Refreshingly, in this age of political correctness, she believes young girls are different from young boys. She was editor of the book, *The Art of Human Computer Interface Design* and in 1991 authored *Computers As Theater*. More recently, she published an online collection of essays on computers, art and nature with the intriguing title of *Severed Heads*. Here to shake up yours on the subject of information technology and culture is Brenda Laurel.

Brenda Laurel

I want to start by telling you a story about my first computer. It was Halloween, and I was almost 12 years old. It was a year when extra-big, everyday things were the most popular sort of costume. (I think this was due to Andy Warhol's influence on popular culture.) Perhaps because we lived in Indiana, my mother came up with the idea of dressing me as an ear of corn. In my grandmother's basement, she fitted me with a chicken wire frame, took tin snips and cut out little arm holes. Then she took cotton cloth and stapled kernels onto the chicken wire; she made crisp green taffeta leaves to camouflage my arms and placed gold-yarn tassels at the top, which tended to fall down over the little eyeholes.

Every year, the local shopping center sponsored a Halloween contest, and the year I was an ear of corn my mother was certain I would win. She had carefully clipped the contest announcement out of the local paper, and when we arrived at the shopping center that evening, she pulled the ear from the trunk and slipped it over my head. The little eyeholes gave me tunnel vision, and I couldn't bend in any direction. I couldn't take very big steps without scraping the skin off my shins with the raw edges of the chicken wire. So at 7:30 we made our way slowly to the tent, a tottering ear of corn and a small, purposeful woman. We arrived only to find it empty. What had happened? My mother dug the clipping out of her purse; she checked the time, she checked the date. She read that the contest organizer was the manager of the Ace Hardware store at the other end of the mall, so she took my hand and dragged me resolutely down the sidewalk as fast as I could waddle. When we got there, I was temporarily blinded by the fluorescent light pouring in the little eyeholes, and my mother dragged me up to the counter. "What happened to the costume contest?" she demanded. I couldn't see the manager, but I could hear his voice. "Well," he said, "we actually finished early. Since everyone showed up at 7, we went ahead." "How could you do that?" my mother asked. "Can't you see that my daughter would have won? And that we were on time and in the right place? What can I tell her now?"

All the little noises in the store went silent. I knew other people were watching. I felt that I'd become an object of wonder and pity. The manager hesitated, and then he proclaimed loudly, "I will give your daughter a prize. She's a wonderful ear of corn." As he led us down the aisle, my leaves swept little hardware store things onto the floor. We stopped at a little toy shelf, and the manager picked up what looked like a gray plastic box with the word ENIAC printed on it. He held it right in front of my little eyeholes to demonstrate how it worked. He showed me a small card with a question printed on it: "What is the distance of the Earth from the sun?" Then he inserted the card into the box, turned a crank and a card came out the other end that read: "92 million, 876 thousand, 479.56." "You see?" he said, "It's a computer. It can answer questions—all kinds of questions. Here are the questions right here," and he held up a little pack of cards that presumably contained all the important questions anyone might ever want to ask. "All you have to do is feed these into the computer," he said.

Well, I had an epiphany. For a moment, I was transported out of my chicken wire cage, out of the world of typewriters and school books and into a glorious age when computers would answer all the really hard questions for us. Perhaps the computer might have even told me what I was doing dressed as an ear of corn in the hardware store. "Thank you," my mother said, graciously. "Next year, please run your contest on time." "I will, ma'am," the manager replied. He shook one of my leaves and said, "I'm sorry honey. You look great, and I hope you enjoy your computer."

My story was set in 1962. America was in its duck-and-cover, Dick-and-Jane period, and the space program was glittering hopefully in the distance. Computers were doing a lot of things that nobody knew about, but the plastic ENIAC—and the hardware store manager's explanation of it—perfectly captured the predominant view of computers in popular culture. This simple techno-optimism was consistent with the idea of progress that characterized the industrial age.

But techno-optimism also contained a poisonous seed of self-doubt. People asked questions, computers bestowed answers. The promise was that we could let them do a whole lot of things for us, but the fear was that if we gave them the power to make decisions, we would lose control over them and, therefore, of our own destiny. After all, computers were blooming as a cultural icon in the deeply anxious period of the cold war. So our fears burst out in cautionary tales like a film called *Colossus, The Forbin Project,* made in 1970 about the defense supercomputers of Russia and the United States. The computers are portrayed as überbeings, with intrinsically amoral natures. They turn against their masters, both good and evil, force them into submission and attempt to take over the world. In this dark view, the shadow dual of techno-optimism has continuously been represented in films since that time; *2001* in 1968, *Westworld* in 1973, *Demon Seed* in 1977, *War Games* in 1983, *Terminator* in 1984. By contrast, *Star Trek,* which was born in 1966, tempered techno-optimism with humanism. Except for the occasional—and often comical—malfunction, Captain Kirk was the master of his computers. In several episodes, he actually caused computers—or robots with computers for brains—to explode by asking them paradoxical questions. So technology was always saving the day, but only with wise, benevolent humans at the helm. This theme also predominates in *Star Wars,* and it continues to thrive in all the permutations of *Star Trek* as well as in contemporary films like *Jurassic Park,* where a Unix-literate girl can save the world from disaster with the help of a graphical interface, and in *Independence Day,* where a nerd with a laptop can defeat the scariest alien technology imaginable.

My point is that the computer serves as a projection surface for our hopes and fears about what it means to be human in these times. It's an important new character in our cultural mythology. Philosopher Joseph Campbell said, "We live in a mythological field, with a hard-wired need and capacity to respond deeply to the symbols that our culture provides." He also cautioned that culture fails us when our symbols are not vitally connected to our lives.

So as a cultural symbol does the computer provide that vital connection to our lives that keeps our culture well? Is the computer a good mythical character? What is the ethos of the computer, its distinct characteristics, its moral nature, its guiding principles? It shows up as an innocuous, zippy little appliance and a world-dominating, soulless megalomaniac—and we recognize it in both roles. It recites nursery rhymes without comprehension, and it locks us out of the pod bay—and we recognize it in both kinds of actions. An athlete hurls a javelin into its Big-Brother eye, and then it pops up again as a little box with a self-effacing smile on its own little screen—and we recognize it in both disguises. Why? Because although it can speak with a human voice or display a human face, we know it is not human. It is a brain in a box without a body or soul, without intuition, passion or morality. It's the last stop on the road to mind–body dualism. It's a severed head—severed from what it means to be human. It's also a megahead, a hypertrophied brain that can become dangerous if it's embodied or self-aware. Indeed, it's different from us, but it's also similar enough that we no longer think of it as a machine or a tool, but as an "other" in relation to the human self. It functions in our cultural mythology to express dualities—mind and body, other and self, logic and compassion, reason and intuition, technology and nature. Yes, we made the computer, but in its role as a cultural symbol, the computer also makes us. So you've got to ask yourself, "What kind of future has this character in it?" And if you don't particularly like the answer, then you have to turn the question around. What kind of character can act out a future that we would like to live in?

My point is that cultural narratives coax technologies into being, and vice versa. That's why the Apollo space program owes a debt to H.G. Wells. It's also why Ronald Reagan's pet defense technology was nicknamed "Star Wars." It's why slick commercials of beautifully dressed housewives effortlessly operating chrome-plated appliances lured women out of the post-World War II workplace. It's why that young pilot in the Gulf War compared his bombing mission to a video game. It's why girls have had such a hard time catching up to boys in computing and why computer software invented by and for females will change both the technology and the ethos of computing.

Culture and technology exist in a dynamic reciprocal relationship. Culture comprehends technology through the means of narratives and myths, and those narratives influence the future shape and purpose of technology. The culture–technology circuit is at the heart of cultural evolution. As we become more capable of radically altering the conditions affecting our biological survival through technological means, cultural evolution becomes the primary factor in our species' ability to survive. The stories we tell quite literally influence our fate as a species.

In his book, *Wonderful Life,* Stephen J. Gould observes that two traits distinguish human beings from their forebears: abstract reasoning and representational art. Tool use was not sufficient to sustain us; other tool-using humanoid species died out. Gould points out that humanity is "an improbable and fragile entity, fortunately successful after precarious beginnings as a small population in Africa and not the predictable end of a global tendency." In other words, we are unlikely. We're not a done deal, nor would evolution predict that we're to be the progenitors of even more intelligent species than ourselves. The story of evolution has neither the unfolding of a divine plan, nor the inevitable march of sentience towards grander and grander manifestations. Rather, extinction is the rule. We are much more likely to die out than transform into a self-aware, infinitely smart, infinitely wise collective shrouded in white light—the Internet, notwithstanding. As evolutionary history would predict, the same set of traits that got us into this mess are the ones that will have to get us out—namely, abstract reasoning and representational art. Computers, it turns out, are extremely good at augmenting both. Coincidence?

When I look into the future, I'm afraid of what I see. I see a world where dominator politics prevail, where human-rights abuses multiply in direct proportion to increasing poverty and overpopulation. I see world religions in a state of rigor mortis, with a death grip on science, art and the exchange of ideas. I see the red rocks of Utah riddled with coal mines and the Amazon a scarred wasteland. Worst of all, I see people in the world who can't talk to each other in any meaningful way. Global networking is a tool of business communications, consumerism, propaganda, banal conversations and mindless entertainment. We all will have forgotten how to tell stories and how to hear them. The majority of the world's population will be very young people without extended families or intact cultures and with fanatical allegiances to dead religions or live dictatorships. We will have what Jonas Salk called "a wisdom deficit"—fewer elders, and even fewer people who listen to them. If we can manifest a different future, we must. The way to accomplish this is to activate the culture–technology circuit—intervention at the level of popular culture.

Many of you are making such conscious interventions, mindfully creating technologies that produce new myths and mindfully making art that influences the shape of technology. It's tricky stuff inserting new material into the cultural organism without activating its immune system. Many artists are content to live on the margin and find their gratification in self-expression. It takes special courage and passion to engage in popular culture. The first example I want to explore is that aging whiz kid—virtual reality. From its strange childhood in government labs and military establishments, VR has emerged as a major concept in the pop-culture scene of the late 1980s. It was hailed as the techno-wave of the future with the potential to transform everything from movies to medical imaging. It was also demonized as the latest in mind-control drugs and the world's baddest war machine. Philosophers adopted it as a platform for renewed debates about the nature of reality, and nearly everyone agreed that a head-mounted display would give you a look inside Pandora's black box.

The mythos of VR reflected our sense of an uncertain future. It became a many-faceted icon for the coming weird times. Why? One key to the power of VR as a symbol is that it was essentially unexpected. It was not a logical successor to the brain in a box. In fact, VR turned computers inside out. The brain-in-the-box computer has no body; VR uses our bodies as its instrument. Rather than presenting framed pictures and pull-down menus, VR gives us first-person, body-centric views. Computers, even today's frisky little portables, immobilize the body in front of a keyboard and screen. Conversely, VR relies upon human movement and kinesthetic sensation to achieve its effect. VR qualifies as what Marshall McLuhan described as an "anti-environment," an inversion that turns the existing environment into an object of attention, criticism and scrutiny.

Turning things inside out is an extremely powerful technique. In fact, inversions have given us some of our greatest leaps in culture and technology and consciousness. In his book *Myths To Live By,* Joseph Campbell describes the change in our consciousness that resulted from our first view of the Earth from space—seeing our blue planet alone in the starry black. Campbell said we suddenly understood that rather than coming into this world, we come out of it, or as Philosopher Alan Watts says, "As a vine grapes, so the Earth peoples." This inversion had a lot to do with the success of the Gaia Hypothesis both as a new area of scientific investigation and as a new popular mythology for our relationship with our planet. My friend and philosopher Terrence McKenna muses about this fascination with turning things inside out. He said that's what made us manifest computers in the first place; it was our attempt to create an instrumentality that could, as he put it, "textualize the world, and exteriorize the soul." VR certainly textualizes the world in the sense that it simulates reality, but the tasks for which it was originally developed, like flight-simulation training and remote operations in nuclear reactors, hardly qualify as exteriorizations of the soul.

Even so, the inversion inherent in VR activated the culture–technology circuit. The public's fancy was engaged, and the time was ripe for some interventions. I was fortunate enough to be involved in one of those—a project called *Placeholder* that my friend Rachel Strickland and I had been cooking up years before the Banff Centre for the Arts and Interval Research Corporation gave us a chance to actually produce it. Along with our colleagues Michael Namark, Rob Tow, John Harrison and a bunch of other crazy people, we created three connected virtual environments, each of which represented an actual place in the local Canadian landscape. The cave was a dark, drippy space, represented as a 3D acoustic environment, which revealed its shapes and textures through sounds; and we represented a stand of hoodoos as a tiled globe of photographic images, and a high waterfall as as full-motion video stretched around a virtual-relief projection screen. Two participants entered the Placeholder together. At first, they could see one another only as tiny points of light that defined both their hands—unlike most virtual worlds in those days where the data glove was a right-handed device. On the walls of the cave were petroglyphs of creatures—a fish, a spider, a snake and a crow—that would beckon you to come closer and talk about themselves. So, for instance, the crow would say, "I am the eye of the world. I see everything that shines and glitters." If you walked over to him and your head intersected with his petroglyph, you would find that you'd become embodied as a crow. The other person in the room would see the crow change color, peel off the wall and begin to move around. If you spoke, you found that your voice sounded like the crow's voice. The best part was that if you flapped your arms, you found that you could fly. We call these animal suits "smart costumes." There were also portals between the worlds in the form of spiral-shaped petroglyphs, and special rocks where you could leave voice messages called "voice marks" in the world. Participants would arrange these rocks and their messages into little story lattices, like audio graffiti. I called this piece an intervention, because we created it with the intention of being able to tell a different story about virtual reality in the hope of influencing the future of the medium through the culture–technology circuit.

At that time, most people were describing VR as an out-of-body experience, and the point of asking human beings to put on the bodies of animals was to bring their attention to the fact that they had bodies in virtual environments. The point of modeling the natural world was to address the common fear that VR would somehow replace the real world for people who became addicted to it. Our intent was to invert this idea, to use virtual reality the way Ansel Adams used photography—to point to beauty and say "notice this," to honor and celebrate the natural world and the ways it articulates our imaginations.

I'm happy to say that this work—although it wasn't a complete realization of our dreams—inspired a lot of talking, thinking and reworking the idea of VR; and it influenced the creation of more works that I would call poetic or spiritual. The best example I know is a piece called *Osmose,* by Char Davies of SoftImage, which was also implemented by John Harrison, the person who worked on *Placeholder.* This piece has been touring for two years, and I think it's the finest work of art created in a VR medium to date. In *Osmose,* one person explores a multilayered world of light, shape and organic forms and almost painfully gorgeous imagery in motion. Char loves scuba diving and for this VR piece, she and her team invented an interface device that uses human breath to navigate in much the same way a diver inhales to rise under water, and exhales to descend. It's amazing what happens to people when they breathe consciously. Everybody looks like they're doing tai chi. People who've never experienced VR before they entered *Osmose* reported to Char that they had transcendental experiences or felt profound peace and joy. One of the construction workers who helped install the piece in Montreal said that after experiencing it, he was no longer afraid of death. Whenever some of these remarkable responses are reported to gatherings of new media artists and critics, Char gets criticized for doing "visionary work," which is, by definition, irrelevant. One critic complained, "It is too beautiful. Why is there not a leaf lying next to a dead body in Bosnia?" How tragic that artists have come to hold the self-marginalizing belief that cynicism is superior to hope. To my mind, there's nothing healthier than rolling up one's sleeves and trying to give the world fresh visions of joy and fresh uses of technology that, indeed, exteriorize the soul. Critics notwithstanding, for the thousands of people whose consciousness was transformed by *Osmose,* Char Davies made a breathgiving innovation.

The last kind of intervention I want to talk about has to do with storytelling. As Bran Ferren pointed out, storytelling is kind of a Holy Grail in the interactivity business. But it's not simply the transmission of narrative material. It's a purposeful action that is intended to communicate, to heal, and to teach. Stories are content, storytelling is relationship. Throughout our history, cultures, families and individual lives have been held together by webs of storytelling relationships. We're all familiar with the grand storytelling that happens at the level of a culture at large; the Homeric myths, classical Greek theater and the plays of Shakespeare are on such a scale. Whether we're talking about myths, plays, novels or movies, these grand stories serve as a means for a culture to behold itself. But they also serve to heal. They present illuminations that help us see where we are, how we got there, what's important and where we should go next. There is no shortage of grand storytellers and grand-storytelling technologies in the 20th century, and I'm sure this will hold true in the 21st.

Grand storytelling fills only part of our need for narrative. There are lots of other kinds of stories floating around our culture today—soap operas, comic strips, romance novels, stories of lost children on milk cartons. Contemporary media, especially television, allows us to feel that we're constantly in touch, always looking through our electronic window at the living heart of our culture. By witnessing what we assume to be a common experience, we have the illusion of relationship. But a talk-show host is not your grandmother, or your aunt or your friend who cares about you. We're vitally interested in the news, even though as cultural theorist Neil Postman observed, "The news generally consists of stories that happen somewhere else, to other people, and about which we can take no meaningful action."

Grand storytelling functions fairly well in broadcast media, but personal storytelling—a tale by an individual that serves to teach or heal a particular group or person—is an endangered art in American popular culture. Over the last five years, I've been involved in a large-scale research project on age, gender and play in American children. As we talked to thousands of children across the country, we learned some things about stories and storytelling. For example, very few of the children said they had ever asked an adult to tell them a story. "Well, did your parents tell you stories anyway?" we asked. "Well, yeah, I suppose while they read me books when I was little," was the typical response. Grandma and grandpa live in another state, or they've passed away, and their stories have gone with them. The art of storytelling still flourishes in certain ethnic groups, and the practice of live storytelling as concert-style entertainment has been gaining popularity in the last five or six years. But for most American schoolchildren, storytelling exists—if at all—as a novelty that occurs in a school assembly.

Now, luckily, some interventions are afoot on the World Wide Web and in multimedia. Artists like Abby Don, who weaves together the family stories of four generations of Jewish women, are an example of how personal storytelling can be enhanced by multimedia. Derrick Powazek's Web site, fray.com, is a gallery for personal stories and a starting place for storytelling relationships.

Here's another story from our research: Recently we've been testing software concepts geared towards girls, one of which has to do with storytelling. We were really surprised that this raised a red flag with some parents. One concerned father in Los Angeles said, "Well, this [product] seems to be a lot about values. Whose values are these?" he wanted to know. "I'm not sure I want a computer teaching my daughter values." Later in the session, we asked that same father how he felt about the values represented in games like *Mortal Kombat* and *Killer Instinct* and, without hesitating, he replied, "Oh, those games don't have any values in them. They're not about values." Isn't that stunning? The values embedded in mainstream video games are so pervasive and so unquestioned that they're virtually invisible to a concerned parent. But if their kid is engaged in imaginative play, with or without a computer, the whole self is exposed. The values category is wide open—trying things on, feeling things out—and if we don't design what goes in there on purpose, we're nevertheless responsible for what falls in there by default.

We humans are perpetually constructing ourselves; that is the central project of the young. We construct our personal identities, our culture, our future and our possibilities, but with what materials? Stories, video games, movies,

Web sites and works of art don't have to be about values to have a profound influence on them. Values are embedded in every aspect of our culture. They're lurking in the very nature of our media and our technology. The question isn't whether they're there or not, but who's taking responsibility for them, how they're being shaped and how they are shaping us and our future. The plastic ENIAC in my Halloween story contained values. The computers, the questions and the answers are all a closed system. All a person gets to do is turn the crank. I think it's time to change that story. The questions and answers are human, and the computer needs to be human too; as human as language, a thumb, a talisman, a fairy tale or a song.

I'm going to leave you with one more story about a computer—this one from the future. There was a little girl walking in the woods. Her grandmother had been in those woods many times. So had her mom, her dad, her older brother, some other people long ago and special friends when she invited them. Some magical beings lived there too. She thought so because she'd seen glimpses of them, and sometimes they left her beautiful secret treasures. The girl went to the woods when she needed quiet, when she needed time to think. Sometimes she would take her friend with her, but there were some places in the woods—her woods—where she would only go alone. There was a certain path she would take when she was feeling sad, a clear little stream she would sit by when she was worried and a waterfall she would visit when she was feeling excited about something. Sometimes she would fly with the owl at night and look at the world through huge golden eyes, and sometimes she would visit the wise salmon who lived at the bottom of a deep, purple pool. She would always find things in the woods; gentle wild-eyed creatures, beautiful stones with stories and secrets inside of them, pictures, messages from her relatives and ancestors, her secret friends and notes that she'd left herself. So she collected these things and took them to a special place in the woods where she arranged them in beautiful patterns. She knew there would always be more things to discover in the woods—her woods—and that someday her own children would walk there too, and they would find her stories, her memories and the patterns of her dreams.

A Half Century of Surprises

James Burke

Our next speaker has asked me to be especially brief, so I will. In 1945, he became head of the Mathematical Laboratory (later the Computer Laboratory) at Cambridge University, where the EDSAC was built. He co-authored the first book on computer programming in 1951 and wrote the first paper on cache memories. In 1980, along with his colleagues, he built the Cambridge Model Distributed System (a pioneer client-server system). Since then, he's worked for Digital Equipment Corporation and the Olivetti and Oracle Research Laboratory. A Fellow of the Royal Society, he has authored publications galore, most notably *Memoirs of a Computer Pioneer,* published in 1985. Please welcome computer pioneer Maurice Wilkes.

Maurice Wilkes

Historians sometimes engage in what has become known as counterfactual history. For example, they might assume that Napoleon won the Battle of Waterloo and start working out what would have happened. Eventually, of course, they'll come up against a check. Another battle is obviously going to take place, and they have no idea which way it will go. Predicting the future is very much like counterfactual history. If you know what is going on in the computer industry, you can make projections into the future—but not very far. You're limited by the impossibility of knowing what original inventions will come along. For example, in the late 1960s, any thoughtful person who noted the growing importance of the local interconnection of computers and peripherals knew it required a better system than the telephone company could offer. He might even have had the insight to see that an improved system would perhaps be based on computer techniques rather than telecommunication techniques. But he could hardly have foreseen the Ethernet appearing when it did. It was a unique and original invention.

Major constraints are placed on what can happen as a result of the laws of nature. As a young man, I came under the influence of G.C. Darwin, who was a very eminent mathematical physicist of the period. He was the grandson of Charles Darwin of evolutionary fame and was an interesting person in many ways. G.C. Darwin became interested in population statistics. He liked to observe that if the population continued to increase at its current rate, which was about 50,000 a day—it's higher now—then by 3000 A.D. there would be room on the Earth for people to stand up but not to lie down. Of course, some catastrophe or epidemic would have slowed the process, or legislators might have passed some sensible laws requiring everybody reaching the age of 60, or thereabouts, to be painlessly put down. It strains our imagination to look 50 years ahead. Darwin was more venturesome. He wrote a book entitled *The Next Million Years*. He was not trying to predict the future. He was trying to say something about what the world would be like, what the pressures and constraints would be. It so happens that the laws of physics are going to impose a serious restraint on the development of silicon chips, something that won't happen in a million years but in the next couple of decades.

Since 1954, the raw speed of computers, as measured by the time taken to do an addition, increased by a factor of 10,000. That means an algorithm that once took 10 minutes to perform can now be done 15 times a second. Students sometimes ask my advice on how to get rich. The best advice I can give them is to dig up some old algorithm that once took forever, program it for a modern workstation, form a startup to market it and then get rich. We're used to doubling speed every two years by shrinkage—making the transistor smaller. But this cannot go on forever. All things being equal, every time you shrink by a factor of two, the number of electrons available to represent a one is reduced by a factor of four. Clearly, this will eventually lead to a shortage of electrons. Obviously, you couldn't get below one electron, but in practice we will get stuck at something more than that—perhaps hundreds. I'm not quite sure. When statistical fluctuations made themselves felt, CMOS would no longer work if you tried to go any smaller. If it did work, it wouldn't be any faster. This is the CMOS endpoint, and we will meet it early next century. It will be the end of the road for CMOS and semiconductor processes in general.

Unfortunately, there is nothing waiting in the wings. Tunneling devices have been disappointing; we have no idea how to build a purely optical computer, and if we did, it's extremely unlikely that it would be any faster than the workstations we have today. So unless something entirely unexpected happens, we must look forward to a period in which circuits will not get much faster. No doubt there will be useful advances, but it will be the end of exponential doubling every two years or less.

Many people think the future lies with a single-electron phenomena in which one electron, more or less, could make the difference between a zero and a one. Such effects have been demonstrated, notably in the Cavendish Laboratory at Cambridge, but only at a very early stage. No devices are even remotely in sight, and its development may well take 50 years.

There is a very painful gap between memory speed and processing speed that is getting worse, not better. There is no fundamental physical reason why a memory as large as a large DRAM (Dynamic Random Access Memory) couldn't be built with the speed of a cache. It would be a one-level physical memory instead of a main memory and a cache memory. I wouldn't make any specific predictions, but I consider it quite possible that there will be a major breakthrough in memory systems, perhaps, one might hope, just when we meet the CMOS endpoint.

There are other possibilities like architectural improvements. But they're getting more sparse because all the wonderful devices invented for speeding up the old, big mainframes have now been incorporated onto chips, and architects are running out of ideas. Many people used to say that we merely had to change over to parallelism and all would go on as before. I used to attract a good deal of flack by pointing out that it was not so simple. It's now generally appreciated that while some programs lend themselves to parallelism, others do not; and I think it's clear that we will be stuck on the older problems, but we'll forge ahead in areas that have not been very prominent in the past.

When resources are scarce, they have to be shared. There once was one computer per company or per university, and users lined up for their turn.

Later, time-sharing systems emerged. Now we do not need to share computer power, unless we choose to do so, and everyone can have their own computer. The important thing is that it is entirely under the user's control; no conflicts with other users. The dramatic improvements in fiber technology are beginning to have remarkable effects on long-haul communications. The time will come when these developments have a big effect on local areas and communications. Specifically, I expect to see more and more dedicated links used to interconnect computers, terminals, printers and so on, instead of local area networks (LANs). The great advantage of a dedicated link is that it is possible to guarantee the quality of service. With a shared link, unless its capacity is divided into slots and one is given to each user, then the quality of service can only be defined statistically. I have an image of computers and servers in a central location, connected by dedicated fibers to terminals and printers. Currently, in fact, you can purchase plug-compatible links that enable you to separate terminals from workstations by a substantial distance. I have one link that was installed while I was away on a trip. When I came back, I carried on as usual, not noticing for a couple of days that the workstation was no longer in my office; it was somewhere else in the building, and then, of course, I realized why the office was so nice and quiet.

This rather excites me. If the terminals in a building are connected to workstations in the offices, and those workstations are withdrawn to a central room leaving the fiber connection in place, then the local area network doesn't disappear; it goes with the workstations and becomes condensed into the central room, giving rise to all sorts of interesting possibilities. Perhaps you could make it cheaper, make it faster or make it into a bus instead of an ordinary network. In fact, you would consider the economy of equipping a building as a whole; providing computers, allowing for the cost of installing the fibers, the cost of racks for packaging, workstation costs and that sort of thing. It's a question of mindset. I doubt if people are willing to go back to the idea of a centralized system, but perhaps when they realize its enormous reliability compared with a LAN of dedicated links, they might think differently.

There are other things I would like to see happen. In particular, I think programming languages took a wrong turn in the late 1950s, and I'd like to see this rectified. It was a great pity that FORTRAN and ALGOL gave rise to two opposing camps when there should have been cross-fertilization between them. There's much that those camps could have done for each other, and we would have had better languages much faster. What I particularly regret is that the people responsible for modern programming languages remained wedded to the ALGOL stack-oriented block structure. One of the reasons for the success of FORTRAN was the absence of block structure. Block structure makes it difficult to include a machine-code subroutine in a program; it inhibits progress in separate compilation and in the reuse of subroutines; and it imposes an unsuitable hierarchical protection model.

I look forward to a time when properly handled objects will be the answer—objects much talked about, not always understood. We need an evangelist to stress their importance. The important thing is that objects can stand alone. They have a clearly defined interface to the outside world. Any routine can communicate with an object if it observes the correct protocol, and objects can communicate with each other; they have no ordering, no pecking order and they're all on one level. As a result of the development of object systems, we are seeing renewed interest in the things I mentioned, particularly the reuse of software.

It has always puzzled me that people are so fascinated by hierarchies. A hierarchical system is good for running an army, for example, and indeed for other control jobs. But if you want to see what can be done with a flat, nonhierarchical system, then you only have to look at the Web. In 1991, I was at a press conference in Turin, which was held at the venerable Academy of Sciences. In the course of its distinguished 200-year history, that academy has acquired many books, and the walls of the room where the press conference was held were lined from floor to ceiling with them. A journalist waved to these books and asked me, "Will all these books be replaced by the computer?" I had no hesitation replying that if the book had been invented after the computer, it would have been acclaimed as a great advance. That served very effectively in getting me off the hook with the journalist, and at that time I could not see that the computer stood any chance of supplanting the book. But my recent experience with the World Wide Web has changed my view. As far as reference material is concerned—not only formal reference works but back issues of scientific journals, for example—I now believe a computer could very well become the principal means of accessing them.

There are, of course, financial problems to be faced, like how to remunerate the author. But since the author of that type of work doesn't get paid very much anyway, it doesn't seem like a great problem. As for the publisher, his production costs would go down very much, and he would need to invest less capital, so we wouldn't want to compensate him very much. But I should compliment my friends in the publishing world by saying that we value their role in persuading authors to set pen to paper.

Then there's the question of quality control. We need protection against homespun reference works filled with misinformation. I don't know whether this is really a problem, but we have to realize that if it becomes easier to spread information, it becomes easier to spread misinformation. As for computers replacing the ordinary type of book—textbooks, novels—I'm less sure.

Fifty years is a long time. People talk about the software crisis, and I'm glad they do because software is the means by which a computer does useful work. If there were no software problems, then computer progress as far as applications were concerned, would end. General Sherman, in his memoirs of the Civil War, writes about the sickening confusion he always found as he approached a battle from the rear. Well, getting results from a computer is a battle, and the software is on the front line. We've made a great deal of progress in software over the years. Anyone who doubts that should note that the software systems we regard as normal today are vast compared to the systems we had 20 years ago. In that sense, we have made great progress, and I think we will make a great deal more.

I do have a despair of operating systems. They desperately need simplifying and streamlining. They allow you to do things in many different ways, one of which would be enough. The designers seem to have no idea of the balance between response time, which is all-important, and facilities provided and so on. I do not confidently predict that operating systems will be any better 50 years hence—and I gather from your cheers that many of you share my concern.

In conclusion, let me share with you a story that's making the rounds at the computer laboratory in Cambridge, England. It turns on the word "exiguous." I don't know about Americans, but most Brits know the word, which means scanty or meager. I don't vouch for its historical accuracy, but the story goes as follows: In the very early days of time sharing, a meeting was held to decide whether a new computer should have the vendor's batch operating system or a newfangled time-sharing system developed at Cambridge University. In the chair was a very senior government official who concluded by summing up the discussion. He observed that the university system appeared to offer many advantages. But he had one doubt, which was whether the documentation was adequate. This question was referred to Roger Needham, who had been responsible for the development of the system. Even in his young days, Roger Needham was very good in committees. He leaned back in his chair, radiated a sense of confidence and well-being and said, "I would say the documentation is exiguous." "Oh, that's all right," said the chairman, "We can go ahead."

ELLIOT SOLOWAY

The Interactive Classroom

12

James Burke

The next speaker earned his Ph.D. in Computer-Information Science at the University of Massachusetts at Amherst in 1978. From 1981 to 1988, he taught at Yale in the departments of Psychology and Computer Science, and since 1988, he's been an associate professor at the University of Michigan both in the School of Education and the Department of Electrical Engineering in Computer Science. He's also professor of Information—libraries, that is—and since his first degree is in philosophy, he's what we might call a well-rounded chap. He holds a number of extremely distinguished teaching awards, and he's published more stuff on artificial intelligence and education than you can shake a mouse at, including a couple of pieces that caught my eye. One was called "How the Nintendo Generation Learns," and another called "Beware of Techies Bearing Gifts." His particular concerns relate to how learners of all ages and stripes will be able to routinely use computational media to achieve their goals over the next decades. He's put his money where his mouth is. Media Text, an award-winning multimedia composition environment used in hundreds of classrooms around the country, was designed and developed in his lab. He's editor and co-founder of *Interactive Learning Environments,* a journal devoted to exploring next-generation technologies for teaching and learning, which is what he's going to talk about. Ladies and gentlemen, to look at the impact of information technology on kindergarten through 12th grade over the next 50 years, here's Elliot Soloway.

Elliot Soloway

You've all gone through education—15, 20 years of it—so in your heart of hearts you probably feel that you know education. You remember the good old days; you figure the system worked for you, and if school were just like that now it would be okay. The other day, Gordon said anything that's cyberizable will be cyberized. Think about it, though. Do you want your kid to have virtual friends in a virtual classroom with a virtual teacher? What if you come home one day and realize that your sixth grader is reading at a second-grade level, but he's a whiz at images and *Photoshop,* he's a whiz with *Hyperstudio* and he's a whiz with *Director.* Are you going to hit the ceiling? What are you going to say to your kid? These are the issues we're going to face over the next 50 years, and we're going to think about your children.

Barney, as you know, is the big purple dinosaur. Kids in the Barney generation—five, six, seven-year-olds like my daughter—think a house comes with a TV set, a bathtub, a telephone and a computer. When a computer crashes, my daughter doesn't take it personally; she starts over. Most adults do take it personally. Most of the Perry Como generation types don't feel comfortable. It's going to be very interesting as my daughter and the other Barney generation kids go through school; the kinds of demands they're going to make on education for the use of technology will be very different from those of the current kids. What happens when kids in the Barney generation become teachers? How are they going to feel about technology?

I'm going to argue three things: First, we need to change the current pedagogy. Right now, teaching is done with didactic instructions, someone who stands up there and tells you stuff. That's not the way we're going to need it in 50 years. Second, without technology we won't be able to implement a pedagogy to help kids develop deep understandings. Third, technology can play a clever, catalytic role in transitioning to that new order.

Even before World War II, our educational system was producing people with good traits for the industrial age—obedience, punctuality and attendance. There were rows in factories; there were rows in classrooms. The kids were not bothering to look around; they looked at the teacher. What was a test in those days? A test was to see if you could repeat back to the teacher what the teacher already knew. Why? Because on the assembly line you don't need imaginative, original, creative thinkers; you need folks who are going to respect authority, who are going to be reliable and who are going to be on time. Paul Simon said it: "When I think back to all the crap I learned in high school, it's a wonder I can think at all." Paul Simon is a success in spite of the schools.

Now we're starting to make the transition to a new kind of order. We're dealing with a process economy, not a product economy. MIT economist Lester Thurow noted that after World War II, all that was necessary were a few entrepreneurs. It was okay to educate only the top 20 percent, because that's where the ideas came from. However, as we make the transition to a process economy where faster, better and cheaper is important, we can no longer afford to educate only the top 20 percent; we have to educate 100 percent. Never before in the history of the world have we attempted to teach and educate 100 percent of the kids.

That is an amazing goal, but if that's what we're going to try to do over the next 50 years, then we need a different kind of pedagogy. Kids need to be able to learn new things on their own, and they have to communicate with their co-workers and their peers. It's hard to disagree with those three goals, but I defy anyone to find them in any state education guidelines in the United States or around the world. What's there? What's the current pedagogy? It's the one your mother knows, that John Dewey knew. My mother used to say to me, "Don't just sit there. Get up and do something." When I got up and did something, I took ownership, I took responsibility, I learned. She also used to yell at me, "Don't just do that. Think about what you just did." That's reflection. All the evidence shows that learners who reflect on what they do are successful learners. The term is called "metacognitive skills." If you step back and reflect on what you've done, then you'll be a successful learner.

At the turn of the century, John Dewey advanced these notions of learning through doing. But you have to be real careful that it's learning through direct experience. You don't want to *learn* about science, which is what we have taught kids up to now; you need to *do* science. Again, we have to be a little careful. It's not just hands-on science—you did that in school. You went into the chemistry lab, and you followed a set of cookbook instructions. All the major organizations—the American Association for the Advancement of Science, the National Research Council, the National Council of Teachers of Mathematics—are all saying the same thing: We have to stop didactic instruction and talk about sustained inquiry whereby kids ask sincere, deep science questions. For instance, "What is the quality of water in the stream behind my school?" That is an authentic, interesting question to the kids. You could teach them all about the different states of matter and different kinds of energy. You could ask, "Where does my school waste energy?" Ah, that's interesting. You could teach about ozone and ask, "Should I wear sunscreen?" Twisting it into a driving question is what changes how kids will learn. It motivates them. All the standards that are coming out indicate the same thing. But it will take upwards of 50 years to implement them in every single school.

What would a sustained-inquiry classroom look like? It wouldn't have rows, and teachers would lecture about five percent of the time as a means of getting people motivated. That's important, but kids in these classrooms would do sustained-inquiry projects for a whole semester. They'd ask the question, "What is the quality of the water in my stream?" For the whole semester, the biology, chemistry, ecology and geology would all be contained in that one driving question.

Of critical importance will be the technology. If you're going to ask kids to do sincere investigations, they're going to need new sets of tools. When we were in school, we all got asked to write a two-page report on such and such. What did we do? We went to the encyclopedia, copied it down and handed it in. Now kids go to the Internet and turn it in. It's the same thing. But if you have to write about the relationship between pH and water quality and why it is the way it is, you need a new set of tools. How are you going to really go deep into a subject if every classroom has 40 copies of the same eight-year-old book? The standards indicate that when kids want to go deep into some area, they'll sacrifice breadth. Given that, we need a more inquiry-based pedagogy, which is straight from John Dewey and straight from what your mother used to say.

What is the role of technology in supporting that? Let's identify some essential problems with education and see how technology might be able to deal with them. There are four elements in an educational system: the school, the community, and two processes—learning and teaching. First, the fundamental issue in a school is how to ensure a healthy, caring relationship between students and teachers. I'm not a Luddite, but in this particular instance, technology is unnecessary. If you look at pictures through the ages, you'll see teachers working with kids, building relationships. *Mr. Holland's Opus* was a tearjerker. Why? Because you could relate to it. If you were lucky, you had a Mr. Holland who looked after you, who took care of you and who looked out for your interests over and over again. The statistics are absolutely clear: When kids have parents or guardians who care about their scholastic interests, they succeed. It's a new term called "social capital." Those kids that don't have an advocate, that don't have a relationship with an adult who cares about them, don't succeed.

Now let's take the student-student relationship. It's about democracy, right? It's about going to school with kids that don't look like you and don't think like you and trying to get along with kids you don't like. The danger with technology is that it will be very easy to create classrooms where the kids all look the same and think the same. In a way, maybe that's more comfortable and easier, but that's not what school is about. School is learning how to get along. Of course, we can use the collaboration technology and the network technology, but we've got to be very, very careful. It's important for kids to rub shoulders with that caring teacher every single day.

The second problem with education is that it doesn't support learning in context. Right now schools are divorced from the real world. Kids think that science is done upstairs in the lab in room 301. They don't see that science has to do with their everyday lives. The *Star Trek* folks had what was called a tricorder, which gathered data when it was pointed at something. In reality, there are E-Mates, little computers that gather data from rivers and streams. No longer are kids forced to do the experiments in the book. Instead, they frame their own problems, create their own experiments and collect their own data. They learn in context.

Similarly, schools are isolated from other community centers, but networking technologies facilitate interesting kinds of mentorships and internships. It sounds trite, but it's true—it takes a whole village to raise a child. You need different folks to be involved. In Detroit, for example, Latino family services are working with inner-city parents who are very concerned about asthma. The schools want to do a curriculum unit on asthma, because it will tie to the state guidelines as well as being of interest to the local community. To bring them together, we're trying to use Web TV to hook up the homes so kids learn that asthmatics need breath meters. Students then use these devices to measure their own lung capacities at home and compare the results with data from other kids in the class. This brings together the home, the community center and the school and gets people involved in those kids' lives.

Well, what about learning? Literacy has to be the core issue, but what kinds of stories do computational media enable you to tell? How does it allow kids to express themselves? Instead of books, kids have tools that allow them to create interactive, dynamic stories by specifying how one character interacts with another. In a make-believe castle, for instance, they can have the dragon slip on a banana peel. What kind of story can be told with this? Should there be words? Consider MIT Media Lab Professor Seymour Papert's observation: "For the next generation or two, one must expect literacy to include some *letter*acy, since our culture's past is so connected with expression through writing. But even if a truly literate person of the future will be expected to know how to read books as well as understand the major trends in art history or philosophy, via whatever other media become available, it will not follow that learning the letters should be the cornerstone of elementary education." Wait a minute. You don't need to read? But let's turn it around. We're successful because we're verbal, right? But some kids are good at mathematics, they're good at chess, they're good at art, but they're not particularly verbal. Why is it that we start with the verbal and then move on? And what happens to these kids? Why can't we start with the nonverbal stuff, let kids get really comfortable with that, let them be successful in creating stories with these interactive kinds of computational media and then move to the verbal? Where is it written that we couldn't do it that way?

What about another aspect of learning: dealing with complexity. All the standards indicate that we can no longer teach kids isolated, decontextualized skills and assume they'll put them together. That's how we learned. Now, if we want to ask kids to build models of complex stream ecosystems, for example, we have to give them new sets of tools. This will involve software and the concept of learner-centered design. Until about 1975, computers had technology-centered interfaces. There was barely enough zorch—power, in computerese—to do the task, so you did whatever the machine wanted. After 1975 the discipline of user-centered design came into being because there was finally enough zorch in the box to do two things: do the application and run the interface. Now that we have plenty of zorch, why can't they be learner-centered? Learners are not just short users. They have unique needs that are different than users. For example, learners grow. By the end of the semester, they've learned something, but the software stays the same. Why can't the software change?

Another point to consider is motivation. The kids play these wonderful video games and all that great stuff, then they look at educational software and they say, "Yuck!" The software must be engaging.

Diversity is also essential. We treat kids as if they're all size 12. They use the same textbooks and read the same pages on the same day, all in lock step.

Each of them is radically different, so why can't the software be very different for different kinds of kids? We have the zorch to run that kind of software; why can't we create it?

The fundamental goal of user-centered design is ease of use. Nothing wrong with that. Learner-centered design has a different goal: to develop expertise. Ease of use is part of it, but not everything. This is a tool that lets kids build models—not just simulations—of complex stream ecosystems. Normally that requires a differential equation. You're not going to get ninth graders or sixth graders to do differential equations. We tell kids, "Spend three years and learn this stuff, and then we'll tell you how to use it." Not a chance.

So what kind of scaffolding can you build into the software to let them create these models. Personalized scaffolding. For example, when they see this program on their computer, kids say, "Oh, that's my stream. I put my feet into that water for a whole year and measured it. I want to know my stream; I want to understand my stream." Now it's not just anybody's picture—it's their own. It anchors the kids and provides them a context in which to understand what they're trying to do. How do they get around differential equations? Well, kids can construct English sentences like, "As the pH goes up, the water quality goes up." Displayed underneath is a differential equation and a graph that represents the information in the sentence. The result is that the kids create fairly complex models. Interestingly enough, without this kind of technology, without this support, without this use, without the software, they couldn't do something—like building models—that's critically important to understanding science.

There's a great quote from Edison: "I believe that the motion picture is destined to revolutionize our educational system, and that in a few years it will supplant largely, if not entirely, the use of the textbook." Not a chance. So if we want to predict whether these kinds of tools will be around in 50 years, what will have to be different? First there has to be availability. Cost is always cited as a problem. Schools will never have enough money to buy computers for all the kids—no matter how many bond issues are passed. The only way it will ever happen is if parents buy computational notebooks for the kids, just the way they're doing it now. This has to be the top priority of our educational system—every kid has to have one. Second, and this is a deeper reason, all the technologies up to now have by and large been one-way technologies—VCR, TV, radio, camera. I'm a professor. I like to tell stories. I don't need technology for that. I'll do it myself, thank you. But I could use help in the classroom. When I'm with 30 kids, and they're all doing different things, they need support. I can't be there with all 30 of them, so I need interactive technologies. That's what's different about today than before; that's why there's a potential for this technology.

Let me wind up with the following: It is very upsetting to see a headline on the cover of a national magazine that reads: "Everybody Hates School." Could this be true? Well, something has gone awry in education and there are all kinds of suggestions on how to fix it. Some even argue that public education is finished; blow up the schools and start afresh; let commercial outlets provide schools. That's really, really wrong. There have to be other ways to do it. Since I'm a techie, I'm going to consider how I can use my technology to find a remedy. People all agree—whether they're liberals or conservatives, Democrats or Republicans—that kids need computers. What we're seeing is people coming to the table and talking about the technology, coming together to make a change. What is interesting is that when they sit down together, they discover that they agree on lots of things, and they want to do something about education. So one way the technology can help is as a catalyst.

Let me conclude by making a plea for public education. In a democracy we have all kinds of organizations, but the cornerstone organization must be public education. It's the place where everybody comes together. They don't look alike, they disagree, but they come together. The idea that we can scrap public education in favor of some other system really is scary, because it attacks the fundamental core of democracy. My parents came to this country in 1946. They didn't speak English; my native language is not English. I stand before you today because I went to a public school, because the teachers and my parents said, "You will succeed." Well, it's payback time, right? Our parents did it for us; our teachers did it for our generation. We've got to do it for the next generation.

REED HUNDT

Delivering Bandwidth Like Pizza

13

James Burke

Our next speaker is a real plus because, with a schedule like his, we didn't expect him to be able to accept the invitation to speak. But he found time, and we are enormously pleased that he did. Today, he takes the broad view—I suppose you'd expect him to since he comes from a legal background—and he's now in public office. He's the first man in his job to develop and expand an Internet site that's now getting 14,000 hits a day. On that site, you can get copies of his organization's decisions, agendas, speeches, public notices and telephone directory. He introduced a fax answer-on-demand system for his clients. He's the organization's first chairman to involve himself in public on-line chat sessions, and the first to have a personal computer on his desk—and to put one on every employee's desk too, making sure that all of them are connected to the Internet. He's also about to give the organization an 800 number. He's committed to making his organization accessible to the general public, he champions the cause of competition and he believes his organization should produce a clear and specific set of rules for implementing policy in the sector of the economy with which it deals. He's a staunch defender of the public interest and for fair rules in business. Now, a lot of that would not surprise you if you were expecting a business tycoon. But if I also tell you that he wants to put the Internet in every classroom by the year 2000, and that he can literally tell you how to make a telephone call, you'll know what I'm talking about. Ladies and gentlemen, here to speak about the particular relevance of telecommunications to the future of the information-technology industry, please welcome the chairman of the FCC, Reed Hundt.

Reed Hundt

1-888-CALL FCC, is our free telephone line, and that is all I want to admit about the actual workings of the telephone system. On Valentine's Day 1996, Vice President Al Gore kicked off your celebration of the 50th anniversary of the invention of the electronic computer. As it happens, the vice president and I were born a year after its invention. My birthday falls on March 3rd, and that was also the birthday of Alexander Graham Bell. How do you think I happened to be selected as the Federal Communications Commission chairman? I want you to know that March 3rd was also the birthday of the inventor of the railroad sleeping car, George Pullman. This was a close call for me. I could have been selected as chairman of the Interstate Commerce Commission, which, in fact, was put out of business by the new Republican Congress in 1996. The Federal Communications Commission they only thought about shutting down.

In any event, being of the same age and growing up in the same town, it came to pass that the vice president and I went to the same high school. There were differences between us; he was slightly younger, he had the prettiest date at the senior prom (a girl with the unusual nickname of Tipper), he was captain of the football team, he was the National Merit semifinalist and he went to Harvard. He also truly loved all science, and if he hadn't gone to Harvard, I believe he might have become a computer wizard. I don't think I said that quite right. It wasn't that he went to Harvard, as Bill Gates explained to me; it was finishing Harvard that was Al's mistake.

Now I've turned 49 slightly earlier than the vice president, and I am holding on for one last spring in my step, one last irrationally exuberant year of what could at least colorably be called youth. Frankly, I'm delighted to note with disdain and ridicule the birthday of anyone or anything on the other side of the great divide, the big 5-0. So, Mr. ENIAC electronic computer, you are a dusty, decrepit, defunct 51, and I laugh at you. A truth of life as you reach your forties is that you are always as young as your ideas. Back home in Washington, the telephone and broadcast lobbyists call my ideas juvenile. However, I should have all the young ideas in the world to cope with the wedding of Bell's progeny and Mr. ENIAC's—also called the convergence of communications and computing. But instead, I've taken the easy way out and brought you questions, not answers.

First, computers have made my three children—age eight, 11, and 14—infinitely more knowledgeable than I was as a kid, and they seem to be getting the topsight described by the brilliant Yale computer scientist David Gelernter. This is the ability, he says, to use the machines to manage and manipulate the world. But my kids have access to computers. What about kids who don't? Is it a problem that the majority of kids in this country—a country in which one out of five children is growing up in poverty—do not have the opportunity to use the amazing things that you invent?

Second, all three of my kids now want to use the single, delightful computer that Michael Dell sent me and millions of other Americans by mail for $2,500 recently. But I can't afford two more—not on my government salary. Where and when can I buy a top-flight, $500 PC, including a monitor and a really fast modem? Maybe the answer to this question is part of the answer to my earlier question about the have-nots.

Third, my kids want faster and more reliable access to the Internet. If we give them policies that will drive massive network-wide bandwidth growth, will that change the way you build the things you build? Will that be part of the answer to the first question?

Fourth, my kids want library-like and TV-like convenience when they get on the Internet. It is easy and entertaining and sometimes edifying to turn pages and channels. By contrast, the Internet's display of pages behind pages is not much more fun than a cluttered desk; and a long scroll seems to roll right out of those tiny little monitors that technology has stuck us with for now.

Don't misunderstand me. I know that I've come to the Rome of the information age. I'm not complaining, but could you please get us big, cheap, better-shaped monitors and a different look to the Internet in general? If that means more of the push of TV instead of the pull of the Web, that's okay with me, as long as my kids and the consumers have the choice of different kinds of pushes. I know these issues may be a little out of your jurisdiction, but as we all tell each other, computing and communications have—are, will be, might be or could be—converged. So I suspect an awful lot of the growth that I'd like to see will come from a better Internet experience.

Bellcore, a name out of the telephone world, just released a study showing that the number of former Internet users equals approximately the number of current users. In other words, for every person using the Internet, there's someone else who tried it and gave up. Mr. Bell did not have this experience with the telephone, and as I recall from my personal history, the only person who ever told me she'd rather write than ever call me again was my date to the same senior prom where Al met Tipper. Henry Ford did not have that sort of abandonment rate with the Model T. "No thanks Henry. I think I'll get back on that horse." This did not happen. Bellcore might not be a neutral arbiter on the question of Internet usage. But we do, in fact, need a big-bandwidth, user-friendly Internet experience and products that meet the real needs of consumers. This is necessary to get that special chemistry that will drive up the penetration rate of computers to that of phones, cars and televisions—2.1 per household. Those are goals that all of us want to achieve.

Network PCs and traditional PCs need to be configured on some assumption about the bandwidth of the networks. At the World Economic Forum in Davos, Switzerland, the leaders of the world, many of whom came directly from the computer industry, said they did not expect now or in the near future to have big-bandwidth networks available to the residential market. The virtual bandwidth that Intel's Chairman of the Board, Andy Grove, talks about as being the model of today's PCs still seems to be their assumption. Bill Gates has said he thinks the telephone companies will do better at providing bandwidth in the future. If I had Bill's perspective on this particular situation, I would want to encourage the traditional telephone networks to strike some kind of understanding of how to market ISDN (Integrated Services Data Network). I can understand why he would have a generous attitude about this.

I read an article in the newspaper about a part-time girlfriend of Mick Jagger who gave an interview about the experience. She said, "I hope I don't end up like him—old and scrawny." This is the way I feel about networks in the United States, and I have part-time experience with our network builders. These are fine companies, but the fact is we don't have big-bandwidth networks; we don't have networks configured to provide high-speed data connections; we don't have that business model seriously at work with respect to the public's switched network. Meanwhile, private industry is seceding from the more perfect union we were trying to develop with communications networks. It's building its own intranets with high-speed data networks and the combination of voice, video and data services that are provided by satellite, microwave hop and private links. In fact, the world does not note this very often, but the greatest piece of commercial spectrum in the U.S. is already allocated for private use by the large corporations of this country, which have more than 5,000 megahertz of spectrum available just for them. These are the oil companies, the delivery businesses and probably many of the companies represented in this room. They have five times the spectrum that we've auctioned, and in our auctions we've had $24 billion worth of business. So if you take out those hand-held computers, you can figure out that we've given a lot of valuable spectrum to private industry. In fact, the law does not even permit us to auction it to those industries because they don't provide the spectrum to customers on a subscription basis, and that was the limit of our auction authority. We've given away all the spectrum to the private sector to build its own private networks, and we're watching private industry build these tremendously rich, big-bandwidth networks.

What about American residences? What about consumers? What about the public's switched network? What about the people of America, who are only going to be reached through subscription services like PCS, Cellular, LNDS and all the other alphabet services we're trying to auction? Maybe they're only going to be reached by the local exchange company and the new entrants, if we ever get any, that really try to get into that market. Aren't we all vitally concerned that the residential users in America should be able to get high-speed data and big bandwidth so they will be able to have all the things they would really want under those circumstances?

I think the answer to our search for a policy paradigm is pizza. If you're sitting at home, you can take out the Yellow Pages, make a couple of phone calls, and you can find five or six places that will deliver small, medium or large pizzas with lots of different toppings, and they'll throw extra things on the side. They will drive out to your house, they will put the pizza right on the table, and you can eat it. That's the way it ought to be with bandwidth. But out there in the real world, nobody wants the change; nobody wants the status quo shaken up so that you could order bandwidth like pizza from your house. The reason they don't is that there's a lot of risk involved in changing the status quo. In its service territories in Michigan, Ameritech now earns two out of every three dollars spent by consumers on communications. That's a pretty big market share. If they get into long distance—and they keep asking me if they can—they won't want their 66 percent market share to go down. They would like it to go up. In the long-distance business, SNET, the telephone company in Connecticut—which got into long distance earlier with a special pass from Congress—has taken 30 percent market share from AT&T, MCI and others in 1996. GTU, which is in 27 states, has targeted high-volume users and has taken one million of the most lucrative customers and long-distance from the encumbrance of AT&T and MCI. So these local exchange companies, which now collect 99 percent of all the revenues spent on communications on a local level, would like to be bigger. We would like them to be bigger; we would like them to be better if that's what the market calls for. We're not trying to pick winners here, but if you want bandwidth delivered like pizza in this country, then you need to have lots of different pizza companies and lots of different bandwidth companies.

If the country wants competition in the communications sector, or in any other sector of the economy, someone has to write the rules to break up the monopolies in these businesses. But not in the old way, by dividing a company into two, three or more different pieces. That's how AT&T was broken up in 1984. I'm not saying it was a bad way, but that is absolutely not what we're talking about right now. The local telephone market is already broken up. There are 1,500 telephone companies in this country, each one a monopoly in its own geographic region. Every one of them accuses us of being too aggressive in writing the rules of competition that allow new entrants to come in. But of these 1,500 telephone companies, each of which believes that its own market has been opened too much and unfairly to competition, not one of them is entering the market of its neighbor. What is going on there? I know that not all 1,500 CEOs are on the same golf course at the same time talking about this business. It is a very daunting challenge to go into a monopolized market and start with nothing—no customers and no network.

Our new idea of competition, and it's a very radical and profoundly good idea, is that the local exchange market will either be shared or bypassed and connected to. Both sharing and bypassing are possible under the 1996 Telecommunications Act. Under this act, if you wanted to start a communications business you'd get the FCC to write a fair rule that would allow you to come in and lease, rent, borrow or share the local loop. Then you'd disconnect that from the switch, and presto! You'd have unswitched ADSL (Asynchronous Digital Subscriber Line) and a chance to market big, deep-dish, large-scale, pizza-size bandwidth.

This has never been tried before and, as usual, the United States is leading the world, which is watching us and thinking, "Boy, maybe this time they'll fail." The United Kingdom is snickering at us because they had a different idea. Margaret Thatcher promoted the cable industry, which would build a separate line directly to the home. With iron determination, Mrs. Thatcher her descendants and the agencies she helped create, excluded all other possibilities. But lately, to their surprise, a wireless competitor called Ionica has arrived with the idea that mobile telephone service doesn't have to be a mobile, fixed, wireless, local loop. Maybe that will give them another tool.

But in our country, we're not picking winners. We're not promoting an industrial policy. We're just saying that anything goes. Lease the switches or the loops and the pieces of the existing network. Then link them up with your own pieces—or build your own separate network—and get a fair interconnection price. This kind of forward-looking incremental pricing is not just arcane economics talk; it is the mantra that is building a global information highway. On February 15, 1997, the United States convinced 69 countries to enter into an agreement through the World Trade Organization—one we had been working on for three years. Each of these countries will set up an independent agency to write the fair rules of competition and, by rule of law, open their markets to new entrants. Foreign investment will be welcomed, not shunned, and all will adopt interconnection policies, allowing connections between networks so that competition can work.

Why did they agree to accept the American idea? Envy, that's the answer. Right now, 16 percent of our gross domestic product is in the information sector; worldwide it is six percent. The difference between the worldwide percentage and our percentage is the job growth in this country that you don't see in Europe. It's the innovation. That envy is what drove everyone to accept our model. I sure hope we got it right, but we're going to give it a good run for the sake of worldwide economic growth.

We're trying to build a domestic information highway and a global information highway because we really believe that this network of networks needs to be public, widespread and ubiquitous. Then it's going to need lots of cars—that's the stuff that *you* invent. These new trade agreements export an idea, they export a philosophy. They open world markets according to principles of law. We're not trading peanuts for peacocks or petunias; we're not swapping exports and imports. This is a 21st century global economy in which no single country's natural resources give it a special competitive advantage. If you've got sand, you can have silicon. We are not saying the information age will be like the industrial age when a nation's special advantage was coal production or the expertise to process iron into steel. We are saying, instead, that everyone can grow together. I sure hope this works because it's desperately important for the globe to build a real, thriving, global economy.

The FCC is supposed to vote, pursuant to the Telecommunications Act, on whether to take $2.25 billion a year from telecom providers in this country and give that money to 115,000 schools so they can install the world's most modern networks in two million classrooms. Those who don't want this to happen are the people who don't want to pay. Some of those naysayers are the telecom providers who wouldn't mind seeing the next guy pick up the tab.

That's very understandable. But if they don't all pay, then it isn't fair. The money needs to be put in a pot and cycled right back into the telecom sector, which will build these networks. It may not get back evenly to the same people who pay in because competition should determine how the money is distributed. The other naysayers are people who don't want government in this country.

I do think we need government. As Abraham Lincoln said, and I'm paraphrasing: The purpose of government is to do what needs doing, but which none of us could do so well acting alone. There isn't anybody here that could bring networks into every classroom in this country. There isn't any business in this country that can or will do that. But acting together with this very modest amount of money—less than one percent of education expenditures in this country and one percent of the telecom revenues—our children can become the primary beneficiaries of the big-bandwidth world. This country spends $88 billion a year on roads. Do you think we could spare $2 billion for the information highway? You know which side of this particular debate I'm on, and I hope you decide to care about this too. This debate will be raging in Washington in the run-up to the 8th of May when we will have a momentous vote.* There will be four commissioners voting—and you can do the math on your computers—but I've been told that three are necessary for a majority. So I would love it if you all would help us out.

---

*The FCC voted to create a pool of $2 billion annually from U.S. telecom providers and distribute it to local school districts around the country based on financial need. That figure was matched by state and local funding, bringing the annual total to $4 billion. In 1998, under pressure from Congress, the FCC cut the pool in half, still making it the second largest public education initiative next to the longstanding federal hot lunch program.

BRUCE STERLING

Weird Futures

14

James Burke

If anyone's going to rattle your cage, our next speaker will. His official biography describes him as an author, journalist, editor and critic. I think you could also say he has no time for people who take themselves too seriously, so watch out. He wrote the *Hacker Crackdown,* a study of electronic crime and civil liberties that led to the creation of the Electronic Frontier Foundation in 1990. He edited *Mirrorshades,* the definitive document of the cyberpunk movement. He writes a pop-science column for the magazine *Fantasy and Science Fiction* and a critic's column for *Science Fiction Eye.* He's also written seven novels, the most recent of which, *Holy Fire,* is absolutely great. He's on the board of directors of the Electronic Frontier Foundation in Austin, Texas, and he serves on the police liaison committee of the local electronics civil liberties group—if anything electronic can be said to be local. Public speaking is one of his hobbies, and he's done it on ABC, CBC, BBC, MTV and many other less-stuffed-shirt places where you do such things. He's appeared in a large number of magazines and belongs to groups too numerous to mention. Ladies and gentlemen, here to talk about matters so mindboggling that you'd better hope he's wrong is Bruce Sterling.

Bruce Sterling

I feel very privileged to be a science fiction writer at this event. It's been a real circus. I've been lucky enough to live in the late 20th century during the digital revolution, and I consider that a great piece of historical good fortune. As 20th century technical revolutions go, I believe this is the best one ever. The computer revolution has been a very sweet, civilized and kindly revolution—almost a velvet revolution. Now I admit that my favorite revolution hasn't been without its dark, damp and sticky areas, but look at the competition: The radio revolution was supposed to abolish borders and unite the world, but after 50 years of radio, Hitler and Goebbels were making horrible use of it; visionaries assured us that air travel would abolish borders and unite the world, but 50 years after the Wright brothers, the sky was full of armed bombers; and as for rockets and nuclear weapons, well, it's been 50 years since their development and yet we've somehow never thrown nuclear rockets across our borders to destroy the world. Atomic Armageddon was always vaporware hype. It never launched. It never became a real product. Just consider that wonderful fact—what a pleasure it is to put the future behind us.

However, it would appear that cyberspace really is abolishing borders and uniting the world in its own subtle, halting, geeky, otherworldly fashion. We have to scratch our heads now and try to imagine what this weird fulfillment of our dreams might really mean. The wise and cynical minority has always expected this revolution to be every bit as awful as the others—maybe even much, much worse. Way back in 1948, the dawn of time as we all know, there was George Orwell's compelling vision of total technical surveillance—Big Brother downloading a boot into your face forever. Orwell was very foresighted, for something very similar is clearly happening today but with the important distinction that we're all Big Brother now. It's not the secret police and the memory hole that controls the Web. The Web grows like bread mold. It's all over the place. If anybody's in trouble from constant, malignant surveillance, it's not the oppressed masses. Who cares about them? Instead, the lacerating pressure of surveillance is on our society's famous people. The real sufferers are the pop stars, the royal family, the sports idols and, of course, our politicians. These poor things. Their divorces and drug habits are piteously exposed, their mistresses or boyfriends appear on cable talk shows, their personal finances are pored over with fine-tooth combs and they never know when they'll be caught on tape or camera. It's enough to try the patience of a saint, and I think it can only get worse.

Pop-culture historians may remember another terrifying computer scenario: the central gimmick of *Colossus, The Forbin Project* in which a malignant, centralized superbrain that knows everything awakens with evil plans to exterminate humanity and fold, spindle and mutilate our souls. To no one's particular surprise, we do have a single, giant, world-spanning cybernetic supersystem now. But instead of a reel-to-reel mainframe, a gloating punchcard menace, it's a touchingly scatterbrained and dotty old thing. Instead of threatening us in a grating monotone, it gently says things like, "URL not found on this server." Today, the darkest and scariest computer-menace scenarios are about information warfare. The very term makes you shudder, doesn't it? As a veteran of the 20th century who knows a little about traditional warfare, I find information warfare to be a rather pale and unconvincing virtuality. If such a war ever breaks out and nobody gets bombed or shot, most of us will turn to the sports pages.

So the computer revolution has been kinder and gentler than we might have expected. One wonders why. I don't think it's because we're any kinder or gentler than our ancestors. My theory is that it's because we never really knew what we were doing. Developments have always been different from what anyone expected. We never expected personal computers, we never expected the Web; we never expected these things, because we never understood what the heck we were getting into with computing machinery. We've always been misled by our preconceptions. We had no idea what computers were 50 years ago, and we rashly called them mechanical brains or thinking machines. In fact, we boldly claimed that these machines were thinking when we didn't know what thinking was ourselves—and we still don't.

Computers have never done any of our thinking; they've only revealed our vast ignorance about the nature of thinking. Frankly, I firmly believe that human brains rarely do much so-called thinking. Somehow we were misled thousands of years ago into believing that real human thinking had something to do with rationality and sequential logic. Human brains do this kind of stuff about as often as human feet do ballroom dancing. When human brains are bubbling along at top capacity, they come up with things like, "My love is like a red, red rose." This makes no objective sense at all, and yet we all know very well what it means. The human brain is all about ideas and meaning, understanding and perception. Computers are not about this stuff at all.

They founder on it because we build them and we founder on it ourselves. Personally, I blame Aristotle, a very glib and persuasive guy who never should have been trusted by anyone. In any case, the next 50 years will allow us to clear away a great deal of confused and antiquated thinking about thinking. We'll stop trying to make computers think, and we'll learn how to make them really compute.

Computation isn't thinking. It's a refreshingly strange, raw and deeply stupid process. It has a lot in store for us. Computation has the power to vanish into the deepest textures of our daily lives. It's a technology with a strong center and a lot of edges. What do I mean by this? Consider television, the center of which is a cathode ray tube that can create visual images from electrical signals. The edge of television is an orange-haired divorcée eating a frozen dinner and watching "Wheel of Fortune." The center of the Internet is the decentralized TCP/IP protocol. The far edge of the Internet may be a world of digital nomads all carrying laptops, cell phones and satellite dishes; congregating instantly and fluidly from every corner of the compass; forming vast, colorful, buzzing tent cities of bubble-packed software and string, and then vanishing like the dew. It's a weird vision, but this is what our physical world would look like if our world somehow adapted the characteristics of modern cyberspace.

Is this really happening to us? Maybe so. Here we are at this exposition from all over the place, and we must have some earthly reason to be here. So are you a jet setter logging frequent-flyer miles, or are you just a techno-gypsy? Can you tell the difference anymore? A lot of people retire into Winnebagos these days. If you're working so hard just to hit the road and enjoy the end of your life, why not cut the process short and leave right now? Settled life was a technical development, too. A settled life was all about a technology called agriculture, about rivers, walls and fertile soil. Maybe five percent of us work in agriculture now. Maybe five percent of us are really tied to the soil. The rest of us are just used to sitting still. If bits really mean more to us than atoms, how many of us will sell off our atoms to live on our bits? I suppose this vision doesn't make much economic sense, but what does? What does economics really mean anymore in an immaterial, so-called information economy? A vast hall of funhouse-mirror sites, all spiders and firewalls, swarming with search engines and copy devices.

We've all been waiting patiently for the commercialization of the Web, for a stable business model for it. But I can't help notice that the Web seems to be growing a lot faster than the economy is. In fact, the Web is growing much faster than any economy has in all of history. We think the Internet may choke up and crash, and maybe it will, but we certainly know that economies choke up and crash. So which system is really more brittle? Which is more artificial? Which is more unreal? Is the Web becoming an aspect of the economy, or is the economy becoming an aspect of the Web? Do we really know what a market is anymore? Do we know what a market might become if many of the major players are automated trading systems rather than human beings? As time creeps on, the edges of the Internet may spawn strange new paramarket entities—truly sophisticated superbarter economies, submarkets, overmarkets, nonmarkets, black markets, gray markets, green markets and hybridized net economies that we have no modern terms for. The way we do business today isn't the natural order. It's not based on holy revelation or bedrock laws of physics; it's just social, political, commercial relationships, far more fluid and vulnerable than we like to admit.

Many people have tried to make 20th century commercial sense of computer networks. We've seen them march in and we've seen them stumble. Prestel, Videotext, Qube, Minitel, the Source, E-World, Prodigy—there will be others. How can there be 40 million people using the Internet, a system that supposedly makes no commercial sense? Maybe when there are 400 million people or four billion people, then we will be forced to admit that our ideas of commerce are no longer sensible.

Networking is permeating the advanced economies now and it's making serious inroads in the developing world. Maybe every human being will have a Net address someday soon, but why stop there? It's just getting interesting. I want Net addresses not just for myself, but for my cell phone and my laptop, my car, my house, my bicycle, my toaster, my surfboard, my wristwatch, my cat and my dog. I think my cat really needs a cellular phone. I'd love to be able to call that cat and know that he was forced to listen to me for once. I want to know what my cat is up to at all times, and if we're wiring our pets for sound, let's network all the important animals such as livestock. Of course, we'll start with prize racehorses and stud bulls, but if the price of the technology crashes, we'll network them all—every sheep, every cow, every goat—millions of them. We already do this for wild animals. We have metal leg bands for migrating birds, radio collars for bears, wolves and whales, but that's just a primitive beginning. If we're spending tens of thousands of taxpayer dollars to save the California condor, then let's treat this as a serious business investment. Let's put full-scale, 24-hour global tracking and video on those birds. The natural world is a precious resource. As 21st century stewards of the rapidly dwindling natural wilderness, we must do our solemn duty by our friends and clients: the world's wild animals. As a good citizen of 2047, I want the wilderness saturated with media coverage.

But why stop with animals? When bandwidth and connectivity become cheap enough, we can saturate the entire surface of the Earth. We might invent a new, all-purpose global net device—a combination weather station, seismometer, pollution sensor, telephone repeater, minibank, parking meter and video–audio Internet telepresence node, all in one. Their time and position would always be known precisely to the last nanosecond and micron; their entire output of signals and data would be recorded and catalogued in real time and they would be as common as traffic lights, fence posts or beer cans. These devices would be especially common where we already have expensive geophysical sensors—around volcanoes, on earthquake fault lines and so forth. But if we're mapping the environmental changes in high-risk areas, why settle for half measures? Why not just map all the environmental changes on the Earth's surface everywhere, all the time? This would surely pay off in the long run.

I suppose that recording the planet's entire surface sounds rather challenging by contemporary technological standards, but the hardware won't be the problem. According to good old Moore's law, the average computer in 2047 is supposed to be 10 billion times more powerful than the average computer today. Clearly, there's a lot of leeway there. Is having the whole Earth saturated with broadcast computation really any odder than having the whole Earth saturated with bad TV sitcoms? Having accomplished this, moving our canned media coverage into outer space would be a relatively minor development, something best left to backyard astronomers and other public-spirited hobbyists. Hauling humans and their life-support systems onto the surfaces of other planets is a daunting technical challenge, but saturating the solar system with tiny machines, each one as smart as 10 billion PCs, sounds like it might be a very pleasant and edifying enterprise in 2047. I'd like to see the moon wrapped in media coverage for our general edification. I'd like to see the sun wrapped in media, too. There aren't a lot of practical hands-on measures we can take with the sun, I guess, but surely it's in the interest of the human race to keep a weather eye on the basic source of all earthly life. We'd do this sort of thing in our spare time if we were more civilized and thought a little more clearly.

Computers don't think. They don't have to think, and we can't make them think. But if computation isn't thought, what is it? It doesn't have much to do with intelligence, but I suspect it has plenty to do with simulation. Computers are very good at creating toy worlds, models, virtualities and artificial spaces. This is turning out to be an arena where computers can really shine. So what can you do with huge, enormously detailed artificial landscapes that have no intelligence? Keep in mind there isn't any will or intention here—just a blind ability to generate complexity and an equally blind ability to sort the products of that computation. Well, that can be very useful. There is such a thing as a blind, nonintelligent design process. It's called evolution. We've already seen some halting attempts along this line in computers, but imagine that times 10 billion. I think it will be quite some time before a computer is able to fully simulate any living creature, even something as simple as a bacterium. However, we do have a pretty good grasp of the grosser laws of large-scale classical mechanical physics—things like jackknives and pipe wrenches. A very powerful mid-21st century computer might be able to fully simulate and evolve simple mechanical objects.

Let's imagine it's 2047, and we want to design, say, a mechanical jack for a car. There are many different kinds of car jacks, so instead of racking our brains designing more of them, perhaps we should just grow them inside computers. Here's the scheme: We'll scan all the existing jacks and input their parameters into our simulation supercomputer. They'll be our seed jacks. Then we'll crossbreed them and torture test them inside the computer for ruggedness, durability and thrifty use of materials. We'll have the computer generate millions of variant car jacks and ruthlessly weed out all that fail. At the end of this process, we'll have our result: an evolved car jack. It's a device that no human brain could imagine—a device produced from methodically exhausting the entire phase space of all possible car jacks. It might not look like a jack at all, but man, can it hold up a car! All its physical specifications are right there inside our supercomputer, so we just cast them out in steel or plastic and behold: an evolving simulation from inside a blind and random computer has become a workaday physical object. It's as real as dirt. You can carry it in your trunk.

This sounds like we're getting something for nothing, and we are. But didn't we get leopards and walruses that way? I don't remember paying anything to obtain the octopus or the pterodactyl, though I could never have invented them from a standing start. If natural selection works in the real world, then unnatural electronics selection ought to work several billion times faster—or maybe it doesn't. Maybe I am simply making all this up. Maybe I'm just inventing it for you, because I like to play with ideas; and for a science fiction writer of my generation, your corner of the world is the place to find them.

Computing machinery has become a spiritual oasis for the imagination. It is a fountain, a mighty wellspring of crazy ideas. The world 50 years from now may have nothing to do with any of these notions, but I can promise one thing: It will have realities, daily realities, physical quotidian realities that are absolutely that strange—probably stranger. It will have things we can't imagine, things we can't make sense of, things beyond our pale, wonderful things. We can't always make sense; human beings die from too much sense. Quite often we're senseless, and we're also fearful. We're paranoid, we live in unreasoning dread and prejudice and our minds are cramped and dark. We are dark and squalid little creatures, and we have huge untapped capacities for malice and wickedness. This is a sordid truth that we hate to admit to ourselves, but we can also survive amazing amounts of surprise and wonder. Maybe that's our saving grace. Sometimes we can overcome our natural, grim suspicion and learn to enjoy a free lunch. Besides, what fun is a revolution that can't promise us a free lunch?

I know I was supposed to get all dark and moody and vitriolic in this speech, but I was just having too good a time. I can't resist this final, golden opportunity granted by our host to give you all a good scolding. So, I'm going to tell you about the darkest and the most morbid and depressing thing that I have actually seen at this event. This happened during an earlier presentation when our speaker expressed his feeling that the idea of leisure is bogus. He polled the audience. He polled you. He asked us, "How many of you can honestly say that you have some leisure?" and one single lonely hand went up. "One guy," he said. Well, I'm proud to report that I was that guy ladies and gentlemen.

I hate to break it to you people, but I'm actually here on vacation. I'm not in your industry. This is something I do for fun—it's a hobby of mine. I've got my feet up and I've got a little paper parasol in my drink. Really, I'm slacking off. Now, if I were working hard and earnestly, I wouldn't be here. I'd be home writing my overdue novel. But instead, I'm here in sunny San Jose getting away from it all. So you see, leisure is not that hard to find in today's society. I'm not even getting paid for this—it's fun.

Now I know you guys have some leisure. At the very least, you have plenty of slack time while you're waiting for *Windows 95* to shut down. Go outside and look at those faces, those photographs of the elite people in your own field. Look at that hall of photographs. You don't see any Scrooges or horrible, shriveled up little drudges over there. Those are obviously multidimensional human beings with deep creative talent who could take some honest pleasure and satisfaction in their lives. So stop pretending that you drudge and labor every working moment. Shame on you for giving in to that evil lie. Workaholism is not a virtue, it is an affliction. If you don't believe me, ask the people who love you. It's not a rational or productive thing to blow out your wrist tendons on the burnout track pretending that you never need any leisure. If you have no time to step back and examine yourself, no time to smell the roses, no time to wonder or ponder or daydream, no freedom to become yourself, I don't care what business you're in, you are collaborating in the theft of your own life.

This is a dark and immoral thing. You will do yourself an injury. But you can get over this syndrome. Hope beckons! Yes, you are brilliant, yes you're very dedicated, yes you're producing maybe 40 percent of the modern American economy, but you're not so deep into this that you'll get the bends if you come up for a little free air. Now I have every confidence in you people. Just put down the mouse and the snack-food bag and take some tottering steps into the full light of day. Eat an apple, go swimming, have children—read a novel, even. Just slack off and glory in your slack! And when you're done, you'll see that living is not drudgery any more than this has been drudgery for me. I am here to tell you that I have seen the future—and it leisures. End of sermon.

RAJ REDDY

Teleportation, Time Travel, and Immortality

15

James Burke

I was going to say that our next speaker is going to take another way out look at things, but having heard Bruce, let's say relatively way out. He earned a doctorate in Computer Science back in 1963, when he came to America from his native India via another degree in Australia. After teaching for a while at Stanford, he moved to Carnegie Mellon, where he was named a professor in 1973. He's now the Herbert A. Simon University Professor of Computer Science and Robotics. And he is recognized worldwide for his work on speech recognition and as the founder of the Carnegie Mellon Robotics Institute, which he ran until he took up his present position as dean of the School of Computer Science there. He's a member of the National Academy of Engineering and the American Academy of Arts and Sciences. He was president of the American Association for Artificial Intelligence from 1987 to 1989. In 1984, he was awarded the French Legion d'Honneur for his work on bringing advanced technology to developing countries, and he was awarded the ACM Turing Award in 1995. His ongoing interest is in human–computer interaction. He has projects running at the moment on speech recognition, multimedia collaboration techniques, just-in-time lectures and automated machine shop. With a background like that, it's all the more interesting that he should choose to talk about something like teleportation, time travel and immortality. I think it promises to tickle the fancy. Ladies and gentlemen, Raj Reddy.

Raj Reddy

As we look forward to the next 50 years, it is interesting to note that when the Association for Computing Machinery was being formed 50 years ago, the sense of excitement was no less palpable than it is today. Vannevar Bush had proposed MEMEX with hyperlinks between documents; Turing, having successfully broken the German code using a special-purpose digital computer, proposed the construction of a universal computing engine called Ace; John von Neumann had recently formalized the idea of a stored-program computer; and Eckert and Mauchly had created ENIAC, the first electronic digital computer in the U.S. There's no question that the last 50 years have been exciting, dramatic and, in many ways, full of unanticipated events which have changed our lives.

What will the next 50 years bring? Given the continuing exponential rate of change, it is reasonable to assume that the next 50 years will be even more dramatic than the last 100 years. When you recall 100 years ago, there were no cars and no highways, no electric utilities, no phone system, no radio or TV and no airplanes, so you can well imagine the magnitude of the change that awaits us!

In this talk, I'd like to share my thoughts on how our dreams about teleportation, time travel and immortality are likely to be realized. One of our most compelling, enduring fantasies of the future has been *Star Trek,* where the themes of teleportation, time travel and immortality have captured the imagination of generations. Will technology make this possible in the next 50 years? We've heard several possible futures in the last two days. I'd like to provide you with one more.

Technology Over The Next 50 Years

By the year 2000, we can expect to see a giga-PC, a billion operations per second, a billion bits of memory and a billion-bit network bandwidth available for less than $2,000. Barring the creation of a cartel or some unforeseen technological barrier, we should see a tera-PC by the year 2015 and a peta-PC by the year 2030—well before 2047.

The question is, what will we do with all this power? How will it affect the way we live and work? Many things will hardly change; our social systems, the food we eat, the clothes we wear and mating rituals will hardly be affected. Others, such as the way we learn, the way we work, the way we interact with each other and the quality and delivery of health care will undergo profound changes. First and foremost, we can hope that Microsoft will use some of this computing power to create computers that never fail and software that never needs rebooting. And yes, I can do without the leisure that I get during the boot time and at the closing time of the *Windows 95,* thank you.

The improvement in secondary memory will be even more dramatic. Many of you know that while the processor and memory technologies have been doubling every 24 months or less, disk densities have been doubling every 18 months or so, leading to a thousandfold improvement every 15 years. Today, you can buy a four-gigabyte disk memory for less than $400. Four gigabytes can be used to store about 10,000 books of 500 pages each—larger than most of our personal libraries at home. By the year 2010, we should be able to buy four terabytes for about the same price. At that cost, each of us can have a personal library of several million books, a lifetime collection of music and a lifetime collection of all our favorite movies thrown in—on our home PC. What we don't have on our PC will be available at the click of the mouse from the universal digital library containing all the authored works of the human race.

If you choose to, you will be able to capture everything you ever said from the time you are born to your last breath in less than a few terabytes. Everything you ever did and experienced can be stored in less than a petabyte. All of this storage will only cost you $100 or less by the year 2025!

How Will All This Power Change The Way We Live And Work?

So how will all this affect our lives? We've heard a number of scenarios for the future in the past few days. I'd like to share some of my dreams on how this technology will be used to save lives, provide education and entertainment on a personalized basis, provide universal access to information and improve the quality of life for the entire human race.

Accident-Avoiding Car

The first invention that will have a major impact on society will be the accident-avoiding car. Let us look at the current state of this technology.

Video of Navlab narrated by Dr. Charles Thorpe

The Carnegie Mellon Navlab Project brings together computer vision, advanced sensors, high-speed processors, planning and control to build robot vehicles that drive themselves on roads and cross-country. The project began in 1984 as part of ARPA's Autonomous Land Vehicle program—the ALV. In the early '80s, most robots were small, slow, indoor vehicles tethered to big computers. The Stanford cart took 15 minutes to map obstacles, plan a path and move each meter. The CMU IMP (Interface Message Processor) and Neptune improved on the cart's top speed, but still moved in short bursts separated by long periods of looking and thinking. In contrast, ARPA's 10-year goals for the ALV were to achieve 80 kilometers per hour on roads, and to travel long distances across open terrain.

With the Terragator, our first outdoor robot at CMU, we began to make fundamental changes in our approach. The Navlab, built in 1986, was our first self-contained test bed. It had room for onboard generators, onboard sensors, onboard computers and, most importantly, onboard graduate students. The next test bed was the Navlab II, an army ambulance HMMWV ("humvee"). It has many of the sensors used on earlier vehicles, plus cameras on pan-tilt mounts and three aligned cameras for trinocular stereo vision. The HMMWV has high ground clearance for driving on rough terrain and a 110-kilometer-per-hour top speed for highway driving. Computer-controlled motors turn the steering wheel and control the brake and throttle.

Perception and planning capabilities have evolved with the vehicles. Alvin is the current main-road-following vision system. Alvin is a neural network, which learns to drive by watching a human driver. Alvin has driven as far as 100 kilometers and at speeds over 110 kilometers per hour. Ranger finds paths through rugged terrain. It takes range images, projects them onto the terrain and builds Cartesian elevation maps. Smartee and D-star find and follow cross-country routes. D-star plans a route using A-star search. As the vehicle drives, Smartee finds obstacles using Geneesha's map, steers the vehicle around them and passes the obstacles to D-star. D-star adds the new obstacles to its global map and replans the optimal path.

Currently, Navlab technology is being applied to highway safety. In a recent trip from Washington, D.C., to San Diego, the Navlab V vision system steered autonomously more than 98 percent of the way. In a driver-warning application, the vision system watches as a person drives and sounds an alarm if the driver falls asleep and the vehicle drifts off the road. The same autonomous navigation capability is a central part of the automated highway system, a project that is building completely automated cars, trucks and buses. Automated vehicles will improve safety, decrease congestion and improve mobility for the elderly and disabled.

Every year, about 40,000 people die in automobile accidents, and the annual repair bill is about $55 billion! Even if this technology helps to eliminate half of these accidents, the savings would pay for all basic research in information technology that has been done since the founding of ACM 50 years ago.

Toward Teleportation

The second area of major potential impact on society is telemedicine. Remote medical consultation is already beginning to improve the quality of care for people located in remote areas. With increased bandwidth and computational capabilities, it will become possible to perform 3D visualization, remote control of microrobotic surgery and other sophisticated procedures. It's not quite teleportation in the classical sense of *Star Trek,* but consider the following: If you can watch the Super Bowl from the vantage point of a quarterback in the midfield, or repair a robot that has fallen down on the surface of Mars or perform telesurgery 3,000 miles away, then you have the functional equivalent of teleportation—bringing the world to us, and bringing us to the world, atoms to bits. Let us look at some recent advances in 3D modeling and multibaseline-in-stereo theory that are essential for being able to do these functions. Can we show this short video please?

Video of 3D modeling narrated by Dr. Takeo Kanade

A real-time, 3D modeling system using multibaseline-stereo theory has been developed by Professor Takeo Kanade and other researchers at Carnegie Mellon University. The virtualized reality studio dome is fully covered by many cameras from all directions. The range or depth of every point in an image was computed using the same multibaseline-stereo algorithm used in the video-rate stereo machine. The scene can be reconstructed with the depth and intensity information by placing a virtual or soft camera from the front, from the left, from the right or from the top or moving the soft camera as the user moves freely. For this baseball scene, we can create a ball's-eye view. A one-on-one basketball scene has also been virtualized from a number of viewpoints.

Currently, this system requires about a teraflop per second for the 3D reconstruction of the basketball scene at the video rate. Instrumenting a football field with a dome consisting of 10,000 high-definition cameras will require 20 petaflops of computation and 100 gigabytes of bandwidth to transmit the 3D model.

Universal Access To Information And Knowledge

Another area that will have a major impact on society will be the creation of a digital library. We already have access to a broad base of information through the Web, but it is less than one percent of all the information that is available in the archives. We can envision the day when all the authored works of the human race will be available to anyone in the world instantaneously. Not just the books, not just the journals or newspapers on demand, but also music, paintings and movies. Once you have music on demand, you can throw away all of your CDs and just use the Web to access anything you want. You may just have to pay five cents each time you listen to it—that could be the way it works. This will, in turn, lead to a flood of information competing for the scarce resource of human attention. With the predictable advances in summarization and abstraction techniques, we should be able to see *Gone With The Wind* in one hour or less, and the Super Bowl in less than a half hour and not miss any of the fun, including the conclusion in real time.

Besides providing entertainment on demand, we can expect the Web to provide learning and education on an individualized basis. The best example of this is demonstrated by the reading tutor, which provides help to students who might otherwise run the risk of growing up illiterate. Can we show the next videotape please?

Video of The Listen Project narrated by Dr. Jack Mostow

Illiteracy costs the United States over $225 billion annually in corporate retraining, industrial accidents and lost competitiveness. If we can reduce illiteracy by just 20 percent, Project Listen could save the nation over $45 billion a year.

At Carnegie Mellon University, Project Listen is taking a novel approach to the problem of illiteracy. We have developed a prototype automated reading coach that listens to a child read aloud and helps when needed. The system is based on the CMU Sphinx II speech-recognition technology. The coach provides a combination of reading and listening in which the child reads wherever possible, and the coach helps wherever necessary—a bit like training wheels on a bicycle.

The coach is designed to emphasize comprehension and ignore minor mistakes, such as false starts or repeated words. When the reader gets stuck, the coach jumps in, enabling the reader to complete the sentence. When the reader misses an important word, the coach rereads the words that led up to it, just like the expert reading teachers whom the coach is modeled after. This context often helps the reader correct the word on the second try. When the reader runs into more difficulty, the coach rereads the sentence to help the reader comprehend it. The coach's ability to listen enables it to detect when and where the reader needs help.

What has been a real plus for the teachers in schools is the fact that children can use it independently. They enjoy reading the stories, and they can prompt the story along. And they're getting some help with individual words that they're struggling with, and they're picking up the meaning of the stories. Experiments to date suggest that it has the potential to reduce children's reading mistakes by a factor of five and enable them to comprehend material at least six months more advanced than they can read on their own.

Toward Time Travel

So this brings us to the prospect of using time travel as an educational tool. In the future, it will no longer be necessary or essential for the teacher and the student to be at the same time and place. Let us see an experiment in which Einstein is talking to today's students.

Video of a synthetic interview created by Dr. Scott Stevens and Dr. Don Marinelli

ALBERT EINSTEIN SYNTHETIC INTERVIEW VIDEO, MARCH 1997
OPENING SCENE
INTERIOR—LARGE CLASSROOM—DAY

ACTION: The large classroom/auditorium is filled with students. They are quiet, intensely watching Dr. Einstein explain the equation E=MC^2. *Einstein is being projected onto a big screen directly from the computer. He is in the middle of his lesson.*

EINSTEIN
The equation E *for energy is equal to* MC *squared. Hmmm. This equation for the equation of mass and energy through the coupling power of light is . . .*

CLOSE UP OF TWO STUDENTS:

ACTION: The two students are seated on the left side of the auditorium. One student turns to the other.

STUDENT # 1
Can you believe that we are actually sitting here taking a class from the great Albert Einstein?

STUDENT # 2
It really is incredible, but I do have one question. Who grades us for this course? I mean, sure, that is Einstein up on the screen, but who is actually going to grade our work?

STUDENT # 1
Have you checked out the teaching assistants?

STUDENT # 1
Teaching assistants? No. Why?

STUDENT # 2
Look!
(He motions to the other side of the classroom.)

CUT TO:
ACTION: Three or four students are leaning up against the wall on the far side of the classroom. Each is dressed exactly like Albert Einstein: wild grey hair, moustaches, lined faces, and each is holding a pipe. They are nodding in agreement with Einstein. The effect should be both funny and amazing.

STUDENT # 1 (V.O.)
Wow!

CUT TO:
INTERIOR—A SMALL CONFERENCE ROOM, PERHAPS AT A LIBRARY—DAY

ACTION: This is a room where individuals can access the computer for the purpose of conducting synthetic interviews. We see a computer terminal on a desk. The computer has a microphone attached to it. There are a few chairs in front of the terminal occupied by elementary school students. They are wearing parochial-school uniforms.

STUDENT # 1
Excuse me Dr. Einstein, we're writing a paper about your life and would like to ask you some questions about your childhood. Could you tell us where you were born?

CUT TO EINSTEIN:

EINSTEIN
I was born on March 14, 1879 in a small town in southern Germany called Ulm. I don't remember it. I remember Munich, whereto my Papa moved the family when I was just one year old. In Munich, my Papa Hermann and his brother, my uncle Jakob, went into business together manufacturing and selling small electrical appliances.

CUT TO:
INTERIOR—SAME COMPUTER ROOM—DAY

YOUNG PROFESSOR
Professor Einstein, why don't you accept and believe in quantum mechanics?

CUT TO EINSTEIN ON SCREEN:

EINSTEIN
Quantum mechanics is very worthy of regard. But an inner voice tells me that this is not the true Jacob. The theory yields much, but it hardly brings us close to the secrets of the Ancient One. In any case, I am convinced that He does not play dice.

I admire to the highest degree the achievement of the younger generation of physicists which goes by the name of quantum mechanics and believe in the deep level of truth of that theory; but I believe that the restriction to statistical laws will be a passing one.

YOUNG PROFESSOR
Well, don't you think the quantum theory is correct?

EINSTEIN
The more success the quantum theory has, the sillier it looks.

CUT TO YOUNGER PHYSICIST:

ACTION: Younger physicist makes face—is exasperated.

CUT TO:
INTERIOR—SAME COMPUTER ROOM—DAY

SENIOR PROFESSOR
Professor Einstein, having escaped from Hitler's Germany, how can you explain the persecution against the Jews?

EINSTEIN
The Nazis saw the Jews as a nonassimilable element that could not be driven into uncritical acceptance, and that

threatened their authority because of its insistence on popular enlightenment of the masses.

CUT TO:
SAME INTERIOR—DAY

ACTION: Two Indian students—a man and a woman—have replaced the professor from the previous scene.

MALE INDIAN STUDENT
Dr. Einstein, I recall reading that you met and became very good friends with Mahatma Gandhi.

FEMALE INDIAN STUDENT
Can you tell me what impressed you most about Mahatma Gandhi?

CUT TO EINSTEIN ON SCREEN:

EINSTEIN
I believe that Gandhi held the most enlightened views of all the political men in our time . . . a man who has confronted the brutality of Europe with the dignity of the simple human being, and thus at all times risen superior.

Generations to come, it may be, will scarcely believe that such a one as this ever in flesh and blood walked upon this Earth.

CUT TO:
SAME INTERIOR—DAY

ACTION: A housewife is now seated at the computer screen.

HOUSEWIFE
Is it true that you never wore socks?

CUT TO EINSTEIN ON SCREEN:

EINSTEIN
When I was young, I found out that the big toe always ends up making a hole in a sock. So, I stopped wearing socks.

END OF VIDEO

So, if we had captured Einstein in living color and 3D when he was alive, it would be technically possible today to have an imaginary conversation with him. The people responsible for this synthetic interview at the Grand Illusion Studios, which is a spin off from Carnegie Mellon, are hoping to create a service which will permit you to converse with your great, great, great grandchildren in the same way. This is not quite the time travel that you've grown to expect from *Star Trek,* but it's another example of substituting bits for atoms to achieve an equivalent experience. With some pre-planning and appropriate data capture, future generations will be able to experience historical events first-hand and interact with past generations.

Toward Immortality

There is work underway in areas such as geriatric robotics that will help senior citizens with simple disabilities lead normal lives well past their prime. And you may ask, can this go on forever? Transplant surgeries are one way of extending life expectancy beyond 100 years or so, and given advances in cloning, we may be getting closer to achieving the dream of immortality. But as Nathan Myhrvold pointed out, you need to download extragenetic experiences—the software in your brain, not just the DNA-based system. One possibility would be to bring you back to life in the fourth millennium using a frozen embryo of your clone and then infusing you with all the experiences you've undergone in this lifetime. Immortality should not be thought of as some mystical transfer of atoms from one brain to the other as in the *Star Trek* movies. It should be viewed from an information-technology perspective whereby you provide the clone with all the important extragenetic experiences of everything you ever said and did. Then you create a rapid, simulated learning environment in which the new clone with a new brain, which can live on for another generation, gets all of your experiences—bits for atoms. It's not quite immortal in the classical sense of the word, but close enough, especially given that the cloning process can go on every millennium. That way you will live forever, except you will be learning the cumulative experiences of all the generations.

In conclusion, the advances of the next 50 years will undoubtedly be as dramatic as the last 50. Capabilities such as accident-avoiding cars, universal access to information and knowledge, entertainment on demand, learning on demand, reading tutors, telemedicine and geriatric robotics will clearly come to pass. More esoteric capabilities such as teleportation, time travel and immortality will also become possible, raising a number of social and ethical questions. As a society, we have to find ways of dealing with these things. As we find ways to transform atoms to bits, that is, substitute information for space, time and matter, many of the constants of our universe will assume a new meaning and will change the way we live and work. This means some of us will have superhuman capabilities, like getting a year's worth of work done in a week. Such capabilities can be used to further increase the gap between the haves and have-nots, or to help the poor, the sick and the illiterate. The choice, I believe, is up to us.

MURRAY GELL-MANN

Pulling Diamonds from the Clay

16

James Burke

By any standards in any field, Murray Gell-Mann is an outstanding world scholar and scientist. He earned a Ph.D. from MIT, like all good and serious people, in 1951. That same year he was already a member of the Institute for Advanced Study at Princeton. The following year he was teaching at the University of Chicago. In 1955, he moved to Caltech where he is Professor Emeritus of Theoretical Physics. In recognition of his unparalleled achievements in one of the fundamentally important fields of science, he has received honorary degrees from many universities including Yale, Cambridge, Columbia, Chicago and even Oxford. He's a director of the MacArthur Foundation, a foreign member of the Royal Society, a former citizen regent of the Smithsonian Institution, a member of the Council on Foreign Relations and a member of the President's Committee of Advisors on Science and Technology. He's also that rarest of breeds, a theoretical physicist whose interests are so wide-ranging—including, to name just a few, natural history, archeology, cultural evolution, historical linguistics, psychology, cognition, population studies, ecology, economic sustainability and geopolitics—that he is as much a philosopher as he is a scientist. Currently, his research center at the Santa Fe Institute—the home of interdisciplinary studies—has been focusing on complex adaptive systems, a matter that brings together virtually every subject of human intellectual endeavor. What gives his work and his imminent talk such extraordinary power is that he has gone beyond the limitations of reductionism to a large, comprehensive overview of our condition. When he talks about computers and what they'll do to us—and for us—over the next 50 years, his canvas is broad. And the really great thing is that his Nobel Prize was for his work in fundamental particles, for his discovery of quarks, and his development of the quantum field theory now known as quantum chromodynamics. It gives me tremendous pleasure to introduce Murray Gell-Mann.

Murray Gell-Mann

We hear, in this dawn of the information age, a great deal of talk about the explosion of information and about new methods for its dissemination. The digital revolution is gathering momentum with new technologies and their applications racing ahead. At least as important as the technical and commercial developments themselves are the alterations that will take place in our thought patterns and our institutions, and that will determine how we respond to those developments and make use of them. ACM97 has offered a wealth of insights into these matters.

It is important to realize, however, that much of what is disseminated is misinformation, badly organized information or irrelevant information. How can humanity extract from that welter of confusing bits knowledge and understanding and even a modicum of wisdom? How can we establish a reward system such that many competing but skillful processors of information, acting as intermediaries, will arise to interpret for us this mass of unorganized, partially false material?

We can easily see that the reward system today is not appropriate. For example, in academic work and in some other walks of life, the principal incentives are for adding little bits of knowledge or understanding at the frontier of science or scholarship. In my field of physics, for example, one well-known experiment may gain someone a chair at a university or at least a promotion to a tenured position, even if the result of that experiment turns out later to be wrong. (Of course, in that case the promotion is not reversed.) But what about clarifying the material in a whole area; synthesizing it, distinguishing—at least in part—the true from the false and the reasonable from the unreasonable, offering the world a clear and reasonably accurate picture of what is understood and what is not? That is often not rewarded to anything like the same degree.

Of course, we must have competing intermediaries. Otherwise, we would be returning to the imposition of authority, a practice that was successfully challenged by European science—then called natural philosophy—in the 17th century. At that time, most European countries organized academies of sciences, for example, that of the Lincei in Italy. In England, the newly restored King Charles II established the Royal Society of London in 1660. Its motto was and still is: *Nullius in verba*—[Don't trust] in anyone's words—not in the words of Aristotle, for instance, but in the comparison of theory with observations of nature.

It is bad enough that in many parts of the contemporary world various breeds of political and religious fundamentalists, as well as old-fashioned tyrants, still impose their authority and suppress dissenting ideas. We who live in free societies must certainly not risk insulating from challenge some orthodoxy in science, scholarship or the arts by relying on too few processors of information.

Fortunately, in the scientific enterprise false orthodoxies are eventually overthrown because science appeals over and over again to observations of nature. Nevertheless, we are all familiar with examples of the persistence of prolonged resistance to correct ideas before they finally triumph. For example, almost all American geologists were hostile to the idea of continental drift until the measurement of sea floor spreading and the accompanying explanation in terms of plate tectonics forced them, as late as the 1960s, to acknowledge that the continents had drifted after all.

The great physical anthropologist Aleš Hrdlička didn't believe that humans occupied the New World earlier than a couple of thousand years B.C. He imposed that view on the scientific community for many years until evidence, such as that from the Folsom and Clovis sites in New Mexico, compelled archaeologists to move the date back to at least 10,000 B.C. Now, of course, still earlier dates are being discussed.

Most historical linguists cling to the notion that family trees for languages cannot legitimately be extended to very large groupings for which the common ancestral tongue dates back earlier than six or seven thousand years ago, because for such ancestral languages, the original sound system is too hard to reconstruct. If those linguists were correct, the evidence for the groupings they acknowledge, such as Indo-European or Uralic, would be marginal. But it is not. The evidence for those groupings, which go back something like six thousand years, is overwhelming. In my opinion, this is another false orthodoxy, one that still holds sway.

In addition to the phenomenon of scientific ideas sometimes yielding slowly and reluctantly to new ones, there is the widespread phenomenon of misconceptions propagating through the media simply through the tendency of commentators to copy one another. For a trivial example, note how many news readers on U.S. radio and television pronounce the "j" in Beijing in the French manner, even though the Mandarin Chinese sound that it represents is a good deal closer to the English pronunciation of "j" than to the French. They are just propagating a "meme" of error through ignorant imitation. Think how often people misuse the quotation from *Hamlet,* "More honor'd in the breach than in th'observance." Referring to drinking wassail, it meant that it was better not to follow the custom than to follow it. But it has come to mean that a certain custom is mostly not followed.

Much more significant are scientific misconceptions that are widely propagated in the same manner. For instance, when a certain kind of subatomic particle disintegrates into two photons, quantum mechanics tells us that whatever kind of polarization measurement is made on one photon, the result also conveys the corresponding information about the other photon. Somehow the erroneous idea is circulating that the measurement of the first photon instantaneously affects the other photon. This is in defiance of the theory of relativity, where such an occurrence would imply that a cause could come after its effect. Of course, nothing of the kind occurs. Instead, a correlation between the two photons has been produced by their common origin in the disintegration that gave rise to them. What is special about quantum mechanics is that this entanglement of the states of the two photons is tighter than would have been possible classically. The result is that any polarization measurement of one of them yields the same kind of information about both. But the common misconception spawns the notion that anything goes in quantum mechanics—instantaneous communication, precognition, psychokinesis, what have you. The U.S. Department of Defense must be flooded with proposals to use the erroneous notion and its crackpot consequences for military purposes.

It is clear that we have a greater need today than ever before for skilled intermediaries who can compete with one another to establish reputations for excellence. But supplying adequate compensation for their efforts presents a problem: Who is to decide which people are worthy processors of material that will in many cases be rather technical? The easiest approach, of course, is to leave the judgment to a marketplace composed largely of laypeople in search of entertainment. But in that way, superficially attractive nonsense may frequently emerge triumphant. We can avoid this phenomenon, which we witness every day on our television screens, only if we make better use of people who make a practice of thinking, knowing and understanding. Perhaps charitable foundations can play a leading role in helping to transform the system of rewards to favor skilled intermediaries who are intelligent, knowledgeable and reasonable as well as successful in appealing to large numbers of information consumers. But this whole issue needs to be studied in great depth and with attention to subtleties.

As has been true for a long time in the established print media, niche markets appear for nonsense like the *Weekly World News.* Remember the headline "Cat Swallows Parrot, Now It Talks: Kitty Wants A Cracker"? The same holds true for somewhat more serious periodicals such as *The New York Times.* Unfortunately, as each of us discovers when we read about our own specialty, even the respected publications get a great many things wrong. Nevertheless, most of us exhibit an amazing gullibility as we tacitly assume that the rest of such a newspaper or magazine is fairly accurate.

One aspect of our craft that is of growing importance is the need to communicate developments in science and scholarship to the public. A century ago it was common for leading scientists, like distinguished figures in the humanities, to write understandably for the educated and interested public about their own work and that of their colleagues. Then, for several decades, such writing was much rarer, and the task of communicating was left largely to science journalists. Good as some of them are, it is heartening to see that nowadays they are joined by considerable numbers of scientists writing their own popular books.

This point serves to illustrate that it is not only full-time intermediaries whose efforts will make a difference; in the long run, it is creative work in the sciences, the humanities, the arts and the professions that will be of greatest help in extracting knowledge and understanding from the immense sea of data that threatens to drown humanity.

The manner in which the digestion and interpretation of so-called information will be handled assumes particular importance as we enter the 21st century—the one in which human civilization must achieve a considerable measure of unity and sustainability if it is ever to do so. We are at the time in history when the curve of total human population as a function of time is finally going through its inflection point—the rate of increase has reached a maximum and is starting down. This is also the time when the human race can produce profound effects on the whole biosphere, whether slowly through economic activity, or rapidly through catastrophic war. As the human race gradually draws together to solve problems that are increasingly global in character—with the aid of rapid communication and dissemination of information worldwide—the associated complexities, contradictions and difficulties loom large on the horizon.

It is highly desirable for humanity to attain unity in diversity. The need for a considerable measure of unity is obvious. However, just as it is crazy to destroy in a few decades a great deal of the biodiversity that has arisen over so many millions of years of biological evolution, so is it crazy to wipe out in a brief period much of the cultural diversity that has been built up over thousands of years of human cultural evolution.

Nevertheless there are contradictions involved. Some traditional cultures, as well as some circles in the most advanced societies, are resistant to the preservation of biological diversity. Some are reluctant to cooperate to solve other global problems. Many are intolerant of the universalizing scientific, secular culture that is sympathetic to democracy and human rights and a principal defender of cultural diversity worldwide. How do we tolerate the intolerant? This kind of dilemma is characteristic of our era. Somehow the human race has to find ways to respect and make use of the great variety of cultural traditions and still resist the threats of disunity, oppression and obscurantism that some of those traditions present from time to time.

The most serious danger is all too familiar: While encouraging the preservation of cultural diversity in the approaching era of the globalization of information, we must be careful not to promote developments that fuel the fires of ethnic hatred, which in many cases is the other side of the cultural-diversity coin. The tendency of human beings to divide themselves into groups that do not get along and sometimes come to blows may turn out to be, in part, an inherited tendency left over from a time tens of thousands of years ago when such behavior may have been adaptive. Certainly, it is not adaptive now, in an age of highly destructive weapons. Also, the tendency of our species to wreak unnecessary havoc on the environment might also be an inherited relic of an earlier age when there were not many people; and the ill effects of environmental abuse, while often very severe even then, were at least geographically restricted.

But even if these tendencies are to some extent inherited, we know they can be modified by culture. That is one of the saving features of humanity. In fact, while most human beings continue to distinguish sharply between "us" and "them," people have made a great deal of progress over the ages in widening the conception of who "we" are. The largest unit of human society to which loyalty is owed has grown from the nuclear or extended family to a clan, a tribe, a city-state, a nation and even a region. A splendid example is the situation in western Europe, where it is now almost inconceivable that a war could break out such as the terrible one that ended 50 years ago. We human beings seem to be moving, although gradually and with many disheartening setbacks, toward supplementing our local and national feelings with a planetary consciousness that embraces the whole of humanity and also, in some measure, the other organisms with which we humans share the biosphere. We hope that many of the efforts to extract meaning from the mass of material circulating in the approaching "global information society" will reinforce this tendency.

Besides the traditional divisions of humanity, we must deal with the multiplicity of groups with common interests, peculiar common beliefs and even common delusions. We have seen how the Internet has made it possible for members of such groups to find one another across geographic and social barriers. Crazy conspiracy theories, new superstitions and urban folk tales flourish and spread as never before. Memes of error are propagated that are much more serious than mispronouncing the name of the Chinese capital or misinterpreting a quotation from *Hamlet*. We see then, that the coming of the information age is reinforcing the simultaneous trends toward globalization and fragmentation that characterize the current era.

But it is not only the fragmentation of humanity into separate cultural entities that creates opportunities, problems and paradoxes for the information age. Just as it is important both to preserve cultural diversity and supplement it with a universalizing planetary consciousness, so too it is critical to foster the specialized fields of science and scholarship and supplement them with integrative work that transcends disciplinary boundaries. Here, reference is made not just to subjects such as biochemistry or geophysics that bridge two neighboring fields, but to research that embraces a great many disciplines at a time.

In order for intermediaries to extract really important knowledge and understanding from the vast amounts of material that will assail us in the future, they must be able to see some of the crucial connections between the different subjects involved—and not just the well-known connections. Intellectual effort must be harnessed to uncover large-scale patterns that are still mostly hidden, emerging syntheses in science and scholarship that are not only important in their own right, but also helpful in making sense of the data that are starting to deluge us. A number of us believe that some of the best opportunities to explore such emerging syntheses lie in mobilizing active cooperation among researchers in a wide variety of subjects. But that kind of cooperation is not so easy to attain at some of the best intellectual institutions.

There are some good reasons why the great research universities and institutes of technology are organized along disciplinary lines, with rather rigid boundaries between fields. Perhaps the most important reason is that our system of measuring excellence is tied to the disciplines. The whole apparatus of departments, degrees, journals, professional societies and sections of granting agencies is based on them. And we know how charlatans tend to cower in the crevices between subjects. Many of us are familiar, for example, with the kind of person whom the mathematicians believe to be a great physicist and the physicists think is a great mathematician.

At leading North American universities, interdisciplinary institutes do exist, but they are usually stepchildren. Headed perhaps by a respected scholar who has earned a reputation in some conventional field, such an institute may be housed in a dilapidated Victorian frame house or a shed built for temporary use during the First World War. The younger researchers have little influence on teaching policy and scant prospects for finding tenured positions. The responsibility for hiring professors and setting curricula is in the hands of the departments.

It is helpful, therefore, to supplement our universities with institutions where collaboration among specialists in a great variety of subjects is more the rule than the exception. One such place is the Santa Fe Institute in New Mexico, which I helped to found and where I now work after spending 40 years at excellent but somewhat more conventional places, such as the University of Chicago and the California Institute of Technology. At SFI, we are not trying to compete with the great teaching and research institutions. In fact, in the course of our work, we hope to help them in two different but related ways: One is to encourage them to move, in their deliberate way, toward facilitating close collaboration in research among faculty members from many different departments; the other is to provide a place where some of their brightest people—who yearn for such collaboration—can engage in it right now.

Most of the research at SFI is connected with one important general area in which numerous disciplines come together: That is the area that I call plectics, the study of simplicity and complexity and of complex adaptive systems—those that learn or adapt or evolve the way living things on Earth do. Complex adaptive systems on this planet include the process of biological evolution, the behavior of individual living organisms and the functioning of certain parts of organisms, such as the human brain or the mammalian immune system. Another complex adaptive system is the human scientific enterprise. Today there exist artificial, computer-based complex adaptive systems such as neural nets and genetic algorithms. (In addition, it should be noted here that the Internet has many of the features of a composite complex adaptive system.)

A significant fraction of the work being done on the similarities and differences among complex adaptive systems is carried out by members of the far-flung SFI family. The specialties represented at the institute include mathematics, computer science, physics, chemistry, immunology, population biology, ecology, evolutionary biology, neurobiology, psychology, cultural anthropology, archaeology, history and economics.

SFI holds seminars and issues research reports on topics such as the AIDS epidemic, the waves of large-scale abandonment of prehistoric pueblos in the southwestern United States, the foraging strategies of ant colonies, how money can be made by utilizing the nonrandom aspects of price fluctuations in financial markets, what happens to ecological communities when an important species is removed, how to program computers to evolve new strategies for playing games and how quantum mechanics leads to the familiar quasiclassical world we see around us.

Those who study complex adaptive systems are beginning to find some useful general principles that seem to characterize all such systems. Some of the important contributions are being made by the handful of scholars and scientists who are transforming themselves from specialists into students of simplicity and complexity or of complex adaptive systems in general.

Success in making that transition is often associated with a certain style of thought. The philosopher F.W.J. von Schelling introduced the distinction between Apollonians, who favor logic, the analytical approach and a dispassionate weighing of evidence; and Dionysians, who lean more toward intuition, synthesis and passion. These traits are sometimes described as correlating very loosely with emphasis on the use of the left and right brain, respectively. But some of us seem to belong to a third category: the Odysseans, who combine the two predilections in their quest for connections among ideas. Such people often feel lonely in conventional institutions, but they find a particularly congenial environment at SFI. They help to overcome the idea, so prevalent in both academic and bureaucratic circles, that the only work worth taking seriously is highly detailed research in a specialty, with more general discussions relegated to cocktail party conversation. While freely acknowledging the continuing centrality of specialized investigations, one should celebrate the equally vital contribution of those who dare to take what I call a "crude look at the whole."

A complex nonlinear system, adaptive or nonadaptive, can usually not be adequately described by dividing it up into subsystems or into various aspects defined beforehand. If those subsystems or those aspects, all in strong interaction with one another, are studied separately, even with great care, the results do not provide a useful picture of the whole. In that sense, there is profound truth in the old adage, "The whole is more than the sum of its parts." But of course the parts, or the separate aspects, are much easier to analyze or model thoroughly in detail, while the theoretical look at the whole must necessarily be comparatively crude. Yet we must swallow our pride and take that look.

We human beings are now confronted with immensely complex ecological and social problems, and we are in urgent need of better ways to deal with them. When we attempt to tackle such difficult problems, we naturally tend to break them up into more manageable pieces. That is a useful practice, but it has serious limitations.

Now the chief of an organization, say a head of government or a CEO, has to behave as if he or she is taking into account all the aspects of a situation, including the interactions among them, which are often strong. It is not so easy, however, for the chief to take a crude look at the whole if everyone else in the organization is concerned only with a partial view.

Even if some people are assigned to look at the big picture, it doesn't always work out. A few months ago, the CEO of a gigantic corporation told me that he had a strategic-planning staff to help him think about the future of the business, but that the members of that staff suffered from three defects: First, they seemed largely disconnected from the rest of the company; second, no one could understand what they said; third, everyone else seemed to hate them. Despite such experiences, it is vitally important that we supplement our specialized studies with serious attempts to take a crude look at the whole.

As far as the problems faced by the human race and the rest of the biosphere are concerned, some of the best studies will, in my opinion, involve ingenious mixtures of scenario writing and computer modeling and simulation. Since much of the research at SFI is carried out by computer modeling and simulation, we are becoming familiar with some of the problems associated with that kind of work. (The advantage is obvious: Computer studies permit the researcher to get a handle on the behavior of models much too complicated to study analytically.) Even if the models and simulations are highly oversimplified, they may share certain mathematical features with phenomena in the real world—those features can then be better understood.

The most important caveat for any very simple model of a complex system is not to take it too seriously. If conclusions from studying the model are to be applied to policy issues, the warning must be even sterner: You may cause great harm if you take it too seriously. On many occasions very evil consequences have ensued from the application of oversimplified scientific—or, even worse, pseudoscientific—models to matters of policy. One of our visitors suggested wisely that such models be used as "prostheses for the imagination." In that role they can be extremely valuable. Sometimes they can give timely warning of serious problems that lie ahead.

A striking feature of computer modeling is the trade-off between detail and transparency. If a model is so general that it can be applied crudely to a wide class of systems, such as those that probably exist on many planets scattered through the universe, then it may be simple enough for the results of a series of runs to be interpreted rather easily. If, however, a few special features of our own planet are built in, chosen from such subjects as terrestrial biology, the nature of human beings or human history and social organization, then the results from running the model will be far harder to interpret, although they may well be more relevant to some real scientific or policy problem.

A way to improve the transparency of a model is to set it up so that many of the assumptions and parameters can be altered at will. Many people can then play with those features in different ways. If conclusions survive that kind of treatment, they are likely to be fairly robust.

One intriguing characteristic of all this work is that the tools of the information age are being used in the search for intellectual syntheses that may help mankind to cope with some of the problems posed by the onset of that age.

Let me close with a remarkably relevant quotation from a sonnet by Edna St. Vincent Millay, supplied by a source that some of you may find surprising—President Clinton's science adviser. The poem is from the book, *Huntsman, What Quarry?*

> *Upon this gifted age, in its dark hour,*
> *Falls from the sky a meteoric shower*
> *Of facts. . . , they lie unquestioned, uncombined.*
>
> *Wisdom enough to leech us of our ill*
> *Is daily spun; but there exists no loom*
> *To weave it into fabric; . . .*

JAMES BURKE

Closing Connections

Conclusion

Let me start, as I said in the beginning, by putting things in context. The fact that two-thirds of the world has never made a phone call, as Reed Hundt noted, puts this esoteric, high-tech interchange into some kind of perspective. There are 5.5 billion people on the planet, but when did you last hear from the other 4.9? Like the kid retorts when his mother tells him to eat his cornflakes because millions are starving around the world: "Name one." With regard to the rest of the world, perhaps the most important thing to be said about this conference on the future is that everybody here was in the future before we started the meeting. It's as though we were living in 2007, roughly the year the average person in the developed countries will be living with the technology we at this conference already have; for most of the old Eastern Bloc, 2015; and for most of the third world, we're already in 2047. So let's keep that in mind when it comes to the direction of the industry and its effect on the world, because the views expressed here put everyone at this conference ahead of everybody out there, and what's out there is much more than just America. If the world described in this highly focused gathering actually comes into existence, we all have to share in helping it do so. As I said before, the future is not already packaged and shrink-wrapped. We all make it day by day in labs all around the country. And the rest of the world is waiting to handle the effects of your work in a future that you are already structuring.

What kind of world did the conference forecast? As one of the speakers said, people involved in technology tend to be optimists—I suppose on the grounds that pessimists jump out of windows and are no longer involved in solving the problems. In general, the participants' view of 2047 was mainly optimistic. Clearly, from a technical point of view, the feeling was that as long as you don't violate the laws—and maybe find new ones—the sky's the limit as far as hardware and software are concerned. The rate of diffusion of the technology is going to continue to accelerate. The costs will continue to fall. The storage capacity, the speed of operations, the amount of stuff on a chip and the processing power of the machines will increase by orders of magnitude. Size will reduce dramatically and ubiquity will become the norm. Beyond that, there's DNA and light to play with.

According to the speakers, 2047 is going to be a world where computers will be invisible, embedded in everything from your bathtub to your brain. In this sense, they will be ubiquitous. Thousands of them will serve each member of society from their places of concealment, making it impossible for anything—or anybody—ever to be forgotten, lost, unrehearsed, disorganized, disfranchised, confused, ignorant, isolated, belligerent, unconnected, inaccessible, unnoticed, marginalized or in need of a good story. It will be a customized world, where anything you want will be manufactured exclusively for you by entrepreneurs that will be changing their identities, their companies' identities, their output and their partners maybe every month to keep up with the constantly shifting marketplace that will gratify your slightest whim at the touch of a button or with a spoken command.

Many people in 2047 will be part of the fast-growing drift from the cities back to the countryside as telepresence makes it unnecessary to live together in a state of mass conformity. The amount of travel in intelligent cars may be drastically less as the same telepresence makes it unnecessary for people to meet in the same physical space, although it's possible that the marginally greater leisure time might put more tourists on the road. But even that might be less than expected given the numbers that will be staying at home, co-

cooned in the latest virtual reality adventure experience—maybe even tapping into other people's life memories.

The world of 2047, according to speakers who touched on this particular area, is likely to be politically fragmented, made up of nation-states, perhaps on the edge of extinction. Regionalism will be rampant, functioning in constantly shifting political alliances, perhaps mirroring what's going on in the commercial world with its capacity for delivering not a customized product but a smart weapon. States that don't own one will use the ever-generous resources of the Internet to find out how to make one. So it will be a world a good deal less monolithic than today's, and for that reason, perhaps a good deal less stable, since the one thing that separatist entities or radical groups will have at their fingertips is the enormously enhanced capability to organize themselves and take action.

In 2047, the general feeling among the speakers, is that the world will be characterized by extreme individualism, with a plethora of electronic servants capable of augmenting the individual's power for self-gratification, self-expression, self-fulfillment—perhaps even self-replication. I prefer Pattie Maes's alternative: agents capable of acting as surrogates for their hosts for everything from political activity to shopping, with the individual perhaps abdicating many of the old-fashioned rights and responsibilities in the process. In listening to all the speakers describe this world, the word that kept coming to mind was "abundance." After millenniums of scarcity of possessions, of information, of access to power, the arts, government, self-advancement, the means of sustenance and comfort, the world of the imagination and storytelling, 2047 looks like a time of plenty—anything you could possibly want. If it's not ready, tell the machine, and the necessary software will be written and the proper action taken before you can say Nathan Myhrvold. So overall, most of the speakers hold a positivist view of the ineluctable progress of science and technology toward 2047.

Back in 1971, *Time* magazine, ACM and others conducted a national computer poll. Those responses were compared with yours, and there were some interesting differences: In 1971, only 58 percent of respondents thought that computers would keep people under surveillance; now 90 percent of you think so. In 1971, only 30 percent thought that computers would help avoid war; today, only 12 percent of you think that. In 1971, 46 percent thought that computers would dehumanize people; with you, that's gone up to 60 percent. In 1971, 38 percent thought of computers as a threat to privacy; 85 percent of you do. In 1971, 49 percent thought computers would improve job security; only 30 percent of you think so. In 1971, 86 percent thought computers would bring more leisure time; only 55 percent of you do.

Overall, the responses also showed that your opinions changed very little through the conference. You obviously came here with a totally open mind. The only significant shifts were that your expectations of job security went down by 10 percent through the conference, and your expectations of electronic democracy went down by the same amount.

As for a few of the offbeat questions, 90 percent of you believe that by 2047 computers will not eliminate the gap between rich and poor, but 90 percent of you believe that they will improve the quality of life. Maybe that means that you think the gap will still be there, but with rich and poor redefined upwards. Eighty percent don't believe that computers will eliminate religion, 90 percent believe that they will not be as intelligent as humans by 2047 and more than 70 percent don't believe that computers will help solve any population problem there might be. Overall, then, the impression is that you are not as optimistic as many of the speakers, and their words did not make you more optimistic. The results may not mean very much on this scale but for the fact that this audience probably has more effect on what's going to happen out there than most.

The mission of this conference was to look ahead to the new century and take the issue into the public arena. So what we've heard could best be described as the beginning of a discussion—a very good beginning. As I said, the people in this room have a head start. For most of you, life is dedicated to making and distributing the technology we've been discussing. There's nothing wrong with that. It's made those of us in the developed nations healthier, wealthier, better fed and clothed and educated than at any time in history, and there is no doubt this historic process will continue.

But there were other considerations this conference helped highlight in the shadowy area between the technologically possible and the effects of its implementation—what you could call the social side of the equation. It seems pretty clear that with regard to implementation—and how to prepare people for the effect of that implementation on society, geopolitics and the biosphere—the various social institutions out there seem to have adopted what might be described as a skyscraper approach to things. They have much the same attitude as the fellow who falls off the top of a skyscraper. As he falls past the 75th floor, somebody calls out to ask him how he's doing. His response after falling past the 55th floor is typical of many institutions and even businesses today: He shrugs his shoulders and says, "So far, so good." That attitude is perfectly understandable, especially if you're living in a specialized world where all you do is look at the bits. In any case, moving outside your frame is fraught with room for error and leaves you vulnerable. Wittgenstein, the great modern philosopher, once described the problem rather well when somebody went up to him and remarked that we must have been a bunch of morons in Europe before Copernicus told us how the solar system worked, how we looked up in the sky and thought we were seeing the sun going around the Earth when, as any kid knows, the Earth goes around the sun. Wittgenstein is said to have replied, "Yeah, yeah, but I wonder what it would have looked like up there if the sun had been going around the Earth," the point being that it would have looked exactly the same. What I think he was saying was that in any decision, you're stuck with what you know at the time. The trouble with that, in the case of our predictions here, is that eventually you have to take some risks because there comes a point where, whether you like it or not, you have to move out of your box to prepare people for the social decisions they're going to have to make about what you're doing. Somebody has to do that. The best people to initiate the necessary discussions may well be those, as Murray Gell-Mann intimated, with some relevant expertise, even if that expertise is highly specialized—people indeed like you.

Social issues don't go away just because you ignore them. Following through with the implementation of the technology requires you to ask some pretty large questions. At the political level, for example, if we use the Internet to dump the system of representative democracy in favor of direct participation, will we get that raw, unfiltered majority that always opts for the death penalty, racial intolerance, short-term, get-rich-quick options and NIMBY tariff barriers that keep jobs at all costs? At the artistic level, will we get a dumbing down as every amateur floods the marketplace with fifth-rate garbage? Do we really want a world of home videos, illiterate scribblings and the kind of rock-bottom, untrained junk that the self-indulgent, talentless, so-called empowerment of the ordinary individual would bring to the arts? In science, how could the average high-school-educated person, with a world of motivation and open access to global databases, take part in the decision whether or not to spend billions of dollars on Murray's research into quantum chromodynamics? In other words, will the new technologies bring with them a welter of mediocrity, vulgarity, mindlessness and self-gratification, the likes of which have never been seen because they had been safely suppressed by technological limitations for all these centuries? Yes, but that's a yes from the years of scarcity when Michelangelo was unique because there was only one of him. What's going to happen to standards when the word itself becomes antiquated, relating only to rarity value in a world without rarity?

What new values will we expose when the knowledge on which they are based is in constant flux? How will each of us decide to behave when there is no longer one right way to do things?

The fact is, this technology is going to bring an unprecedented rate of change to a centralized society not ready for it; a society trained by millenniums of monolithic social structures generated by extremely limited forms of technology that offer virtually no freedom of action; a society shaped by the view that people need leaders to do their thinking for them; that the most valuable knowledge can only be specialized knowledge and that the only measure of intelligence worth a damn is a person's ability to pass tests of literacy, logic, numeracy and, above all, memory; that the best way for people to fulfill themselves and lead satisfying and productive lives is to go to a place of mass work from 9 to 5 every day and define themselves by what job they do. The consumerist use-it-or-lose-it approach to resources may have helped bring the biosphere close to ruin.

Socially speaking, the next few decades are going to make everything that came before look like "See Spot run." We don't know what new forms of social and business structures technology will bring because those human brains out there—increasingly networked, better educated and more productive—may come up with secondary-level social applications that we never even thought of. But on the basis of what we've heard here, what's coming can be characterized rather generally as a move away from the way it's been through history so far—rules and regulations, everybody knowing where they stand, how to behave, what to expect—toward a kind of crazy free-for-all. Obviously, the key question is, how free?

The problem is that ever since the first stone ax, society has had to face the problem of innovation taking place faster than institutions can keep up. Private entrepreneurial businesses always drive change, fueled by a highly competitive marketplace in which most social institutions are at a loss. Social institutions don't seem to be terribly capable of seeing much beyond tactical intent, and yet we need these institutions if we are to maintain stability.

That is why, in a sense, this conference has just begun, and not just ended. The discussion we've had is the opening engagement to spotlight the issue of how to prepare society for 2047–a process for which you may have already laid the foundations. If nobody attempts to widen the discussion to include the rest of the community, then, like the guy falling off the skyscraper, we are ignoring the significance of the approaching sidewalk. We urgently need to get everything we talked about here—and much more—into the public arena. Maybe it should be taken there, as Murray suggested, by qualified and articulate intermediaries—maybe you. There needs to be much more research—by the industry and by the social institutions—on the entire issue of the social processes and how you will change them. And nowhere is it more important, as Elliot Soloway so eloquently said, than in the field of education, because today's schools are where the future becomes bright or dark. Those who will change society with your technology are now sitting in those schools, oblivious to everything that has been said at this meeting. Whether we like it or not, these kids will be presidents, prime ministers and leaders in 2047, and nothing can stop that from happening. How and what they learn today will shape their use of power in 2047—and shape the world.

In 1812, something Thomas Jefferson said sums up this conference much better than I have. He said, "I am not an advocate for frequent changes in laws and constitutions, but laws and institutions must go hand in hand with the progress of the human mind. As that becomes more developed, more enlightened; as new discoveries are made, and new truths are discovered; and manners and opinions change with the advance of circumstances, institutions must advance to keep pace with the times."

DAVID J. KASIK

The Relics of '97

James Burke

David J. Kasik was the exposition director for ACM97. He is a Technical Fellow at the Boeing Company in Seattle, where he is the architect for information systems software engineering, geometry and visualization, and user-interface technology. He designed, developed and published the formal architectural strategy and implementation for user-interface management systems in the early 1980s. His primary focus is in user-interface and 3D computer-graphics technology, an interest that began in the late 1960s with his work on both interactive satellite graphics sys-

Acknowledgments: The ACM97 Exposition was made possible with the help of generous underwriters (Hewlett-Packard, Intel, Microsoft and Sun) and sponsors (Cadmus Journal Services, IBM, Netscape Communications, *Popular Science,* Sheridan Printing, Silicon Graphics, Softbank and Unisys). The contributors provided much of the best computing technology in the world.

A special thanks go to the volunteers and staff who did an extraordinary job of pulling the event and this chapter together: Mary Axelson, Linda Branagan, Mike Carrabine, Coco Conn, John Dill, Clark Dodsworth, Paul Graller, Jim Hoard, Haim Levkowitz, Ken Maas, Jim Martin, Jeff Mayer, Fraser McClellan, Marshall Pittman, Patric Prince, Dana Rennie, Jim Rhyne, Dan Ringler, Ed Silver, Gurminder Singh, Mitch Somers, Dave Spoelstra and Mike Weil.

tems and computer animation. He earned an M.S. in Computer Science from the University of Colorado in 1972 and a B.A. in Quantitative Studies from The Johns Hopkins University in 1970.

Kasik has published numerous technical papers and taught many courses on topics that include systems-engineering architectures, graphics-application development, three-dimensional computer graphics, user-interface management systems and software testing. He served as exhibits chair for the Special Interest Group in Computer Graphics (SIGGRAPH) annual conferences in 1980 and 1994; and from 1981 to 1992, he acted as a liaison between SIGGRAPH and the exhibitor community. Kasik has been an active member of the ACM and SIGGRAPH for more than 25 years.

David J. Kasik

The Relics of '97

As part of ACM97, an exhibit was constructed at the San Jose Convention Center to illustrate what scientists in 2047 might have found in Silicon Valley. It was set up as an archeological dig site—the idea being that a Global Blackout had buried the artifacts of the nascent computing age for 50 years. With a nod to all Sci-Fi B-movies, here was my journey to the site:

> We entered the dimly lit space with the nervous anticipation of children exploring a secret cave. Jagged rock faces straddled the ramp leading to the central dig site. A 1950-vintage digital flight computer lay lifeless in the once fertile Silicon Valley soil. This was the opportunity of a lifetime—a chance to investigate the greatest find in the quest to understand the technology of the Paleotechnic Era—the period when digital technology was thought to have bloomed.
>
> For years, scientists had desperately searched for clues to piece together this era, but the Great Blackout had wiped out all information. Apparently, the global outage was triggered by a massive shutdown of networks and power systems—an earthquake? a computer bug?—and Silicon Valley was hit particularly hard. It was a real tragedy, and everything was lost.
>
> No one knew exactly what had caused this catastrophic event and the subsequent demise of the Paleotechnic civilization, but at least now we had found the relics that would enable us to reconstruct it. The best nanotechnicians huddled around large, clunky

> machines with strange interface devices and display technologies. It seemed that early computing was done with silicon chips in stand-alone machines as large as a meter across!
>
> Early evidence from the site suggested that a digital revolution began shortly before the millennium. It was apparently an era steeped in technological optimism: A passion for the Internet was quickly unifying people around the world, overcoming years of fragmented and provincial communication; dictatorships that relied on misinformation to boost their power were countered by a world with immediate access to a wealth of contrary and contradictory information; individuals used computers to bridge cultural and geographical barriers within society itself.

Until its ultimate demise, the Paleotechnic civilization was giddy with its own potential. Laying here before us were examples of how these new breakthroughs in computer technology were being creatively applied—from the arts and entertainment to science and medicine. This is what our journey revealed:

The Arts

As we proceeded through the site, it was apparent that computer technology at the end of the millennium was already being used to augment human creativity. We stumbled upon a project from the University of Illinois—Chicago that allowed individuals and groups to create three-dimensional computer-based music and sound that could be felt as well as heard. *Vibescape II* was the virtual music/sound instrument that participants played to create virtual compositions, manipulate sound and perform with each other. In real time, it produced 3D audio, giving the illusion of holographic sound that composers felt while lying, sitting or standing on a device called a *VibraFloor.*

Human motion was becoming a standard input to create computer animation because animators realized it was easier to capture motion than model it. Strategically placed sensors on live actors tracked their movements. The Japanese ATR Media Integration and Communications Research Laboratories had created *Virtual Kabuki Theater.* A system of cameras detected precisely where a person was and used the positioning information to create a three-dimensional entity in cyberspace complete with a full range of facial expressions.

For artists and animators, a program developed by Alias|Wavefront used highly efficient interfaces to improve a person's ability to create models and animation sequences. Other, more traditional environments were being developed for graphic artists and industrial designers. The goal was to provide a large-screen interactive table where the physical art of drawing was more consistent with the skills learned in art school and to facilitate group collaboration.

Computers were providing an excellent mechanism to recreate and modify works of the great artists past and present: The *Artificial Artist* from the MIT Media Lab captured the general style of Alexander Calder's sculptures and allowed users to design portraits in steel—kinetic mobiles that expressed their energy and drama. Clearly, people alive at the time foresaw a future generation of computers that would have an eye for design and apply their aesthetic and creative judgment in art and architecture.

Leisure

Other relics indicated that the Paleotechnic people cherished their leisure time as much as any civilization. Computers then—with their exponential growth in processing power—were allowing users to explore new ways of spending their free time. You would think this would have gotten people out of the house more. But curiously, most seemed to prefer to spend lots of time in front of their clunky machines playing violent, software-based games! Matsushita Electric Works even built virtual reality into a conventional stationary bicycle, a massage chair and a virtual horseback ride to encourage people to exercise longer.

Fortunately, chess was still the game of choice for many, which led to the exploration of new ways of representing human intelligence and strategy. The IBM TJ Watson Research Laboratories developed a system called *Deep Blue* that had defeated the human international chess champion. Evidence suggested that immediately after its victory, *Deep Blue* demanded stock options and a company car!

Cybercafés, like the one we found built by Hewlett-Packard, were all the rage, with the brew acting as fuel for those playing the old Nintendo games as well as those accessing the Internet. It was evident that the goal of Paleotechnic companies was to integrate computing technology into leisure activities at an affordable cost. As computing technology seeped further into everyday lives, some people tried to escape by going to places that seemed to be computer-free—though they really weren't.

Nomadic Research Laboratories developed *Behemoth,* a bicycle for those who wanted to exercise while they stayed connected with the world of computing. *Behemoth* was loaded with enough computing and communications equipment to allow people to ride in the comfort of their living rooms, yet simultaneously perform personal-computer tasks, surf the net, navigate, water the plants and any other activity of their choosing.

Education

Life wasn't all fun and games for the pre-Blackout masses. They were able to appreciate the computer for its potential as a learning tool. In the home, computers were becoming user-friendly enough to make families feel comfortable with advanced applications. Smaller children were often more comfortable than adults of that time who didn't have the benefit of growing up on personal computers. With the wealth of educational software available then, learning was fast becoming one of the cornerstones of quality family time. The area we found that collected examples of the best educational programs was impressive even by 2047 standards.

In the classroom, teachers were beginning to use any technological means possible to communicate concepts, manage instructional tasks and allow students to "see" theory. They used a wide variety of techniques to enhance the learning experience: Games were used to reinforce sophisticated concepts in a child-friendly manner. The Institute of System Science at National University of Singapore created a three-dimensional virtual society for children called *KidSpace*. Modeled on Singapore in the 1870s, it was targeted for children ages seven through 11. It asked them to cooperatively develop and manage the social and economic aspects of their community. Players could choose professions; develop, accumulate and trade resources and then use those resources to extend the social and physical environment. By working together to cope with limited resources, small populations, adverse conditions and a strategically important location, players acquired hands-on experience dealing with the issues central to Singapore in the 1870s. (Perhaps this is the reason why in 2047 there are so many urban planners under the age of 12.)

The software makers also had the good sense in 1997 to realize that the best teachers of children were children themselves. The UCLA *Kids Interactive Design Studio* provided such an environment. This program allowed children to develop software for other kids, which turned out to be an effective way of teaching them about technology and at the same time supported learning for others.

Educators were continually searching for innovative ways to teach abstract concepts in science and mathematics. Computer animation and interactive systems had been used to convey the subtleties of mathematical theory as well as scientific phenomena from astronomy through genetic manipulation. Project *ScienceSpace* from the University of Houston used virtual reality to immerse students in a three-dimensional visual and acoustic environment where they could manipulate and master the scientific concepts they were studying.

In fact, virtual technology was providing an alternative to the traditional classroom. The *Virtual University* from Canada's Simon Fraser University had been developing telelearning, which gave students the opportunity to experience advanced learning models with "courseware" that was developed for engineering, social sciences and the arts. Boeing showed a variety of computer-based techniques to teach people how to fly and maintain commercial airplanes. Outside the classroom, technicians wore portable computers to give them on-the-job instructions to address complex maintenance tasks.

Medicine

In the late 20th century, the priority was to improve health care while keeping costs low. Doctors were becoming increasingly reliant on computers for quicker, more precise—and therefore more cost-effective—diagnoses. The World Wide Web was becoming an integral part of the medical community as a tool for distributing medical knowledge and a means by which doctors could assist in medical procedures from anywhere in the world. The East Carolina University *Tell 'Em Edison* project showed how doctors in North Carolina could diagnose strange Silicon Valley syndromes in real time.

In operating rooms, surgeons were benefitting greatly from computer-assisted medicine: Real-time computer graphics were aiding in diagnoses, surgical planning and ultimately helping doctors prescribe the most effective post-surgical therapy. Surgeon-controlled image guidance was assisting in a wide variety of cranial and laparoscopic procedures to minimize surgical invasion. Computer-guided robots developed at Johns Hopkins University were used to assist during spinal surgery and fabricate custom-built artificial hips.

Computer-generated models of human anatomy were also improving doctors' abilities to prescribe medical solutions for those with physical disabilities. The University of California, Berkeley, had developed *OPTICAL,* an acronym for "optics and topography involving the cornea and lens." Computational geometry algorithms were used to fit a precise surface model to points measured directly from a patient's eye. The resulting model was then used to help reconstruct the shape of the cornea. As a consequence, fitting contact lenses became more precise as did corneal refractive surgeries. The *Archimedes Project* at Stanford University developed a number of techniques to supplement physical handicaps. Stanford and LC Technologies, for example, were working with eye-tracking to help those without the use of their hands communicate via computers. A small camera followed eye movement on the screen. Not only could users select commands to synthesize speech, access the Internet and use the telephone, they could also use an on-screen keyboard to visually "type" up to 20 words per minute.

Working

In the workplace, virtually every job was being examined to determine how computers could improve efficiency and productivity. And because the global economy was expanding quickly at the end of the millennium, the focus of corporate computer technology was on linking major multinational projects. IBM was one of the leaders in enhancing productivity by making it possible for multiple users to hold meetings without leaving their offices. The use of computer technology in designing offices was also becoming more popular: IBM was embedding computers into chairs, desks and even clothes to provide easy access to information and resources. Microsoft was working with voice- and gesture-activated interfaces, making for faster transmission of information than through the use of conventional keyboards.

For journalists, small portable computers were replacing the notepad and tape recorder as the state-of-the-art means of news gathering, allowing for the almost instantaneous filing of a news story from the field.

Construction workers relied on instructions delivered from systems like *ARC* (Augmented Reality for Construction) developed at Columbia University. *ARC* was a prototype system for assisting in the assembly of a spaceframe structure made from nodes and struts. A head-worn display presented graphics and audio instructions that guided the worker through the step-by-step process of actually assembling a full-scale aluminum space frame. Two-dimensional versions of augmented reality were already being used to build wire bundles in the aerospace industry.

Modeling the World

Most of the growth in computer technology was in response to solving existing problems. But the main force behind technological advances in the computer sciences was modeling, which was becoming increasingly important in every profession as a means of working out the bugs of a project before any action was taken. Two inherent tendencies of modeling were pushing the advances in computing technology: First, the size of the models themselves were growing at a lightening pace, which increased the demand for better computer performance. Second, models were used to represent objects that were physically too large or small to be easily comprehended without computer assistance.

The Earth was one of the large models that was mapped and measured in numerous ways. A German company called ART+COM integrated data from high-resolution 2D satellite images and 3D altitude data, customized networked algorithms and databases and high-performance graphics. The resulting system, called *T_Vision,* was an experiment that used the whole Earth as its primary and familiar interface to let users organize, access and visualize any relevant information about the planet. By manipulating a large globe (the input device), users could begin with a view from outer space and smoothly work their way into the interior of specific buildings. The developers planned to use the results of the experiment to build better ways for people to sort through the masses of information now at their fingertips.

Architectural planners and designers were using computer-generated models to better understand how new projects would affect the environment. UCLA had developed a method of analyzing large sections of the Los Angeles cityscape in the *Virtual Los Angeles* project. The models included detail as fine as the graffiti on building walls. Planners were already using *Virtual Los Angeles* in the construction of a new UCLA medical center. Virtual reality was also brought into the mix. Silicon Graphics produced *Virtual Reality Theater*, which allowed a group of people to immerse themselves in real-time views of complex terrain. This allowed users to gain an accurate feeling about the real impact of new and futuristic designs.

Generating complex models for computers required users to understand splines, spheres, cones, free-form surfaces and other mathematical objects. *Sketch* from Brown University showed how a computer could infer a complex geometric shape with more conventional free-form drawing techniques.

A Taiwanese research institute, ITRI, was focusing on doing more with cheaper computers using a software program called *IMVR* (Image-Based Virtual Reality). *IMVR* worked to achieve the proper balance between the size and complexity of models and allowed users to display realistic virtual environments and 3D objects in real time on most personal computers without any additional hardware acceleration.

Paleotechnicians seemed to have the most difficulty in producing models that represented human knowledge. The Air Force Institute of Technology had combined probabilities, expert systems, knowledge, and inference into *PESKI*, an intelligent system that assisted in knowledge capture. The system could take a concept and, with the help of multiple agents, make precise decisions about how best to capture it in a computer information base. Knowledge capture became increasingly important as people wanted to understand design decisions for increasingly complex systems like automobiles, aircraft and even computers.

One of the practical results of computer-generated modeling was that users could explore all the possibilities of a potential project—parameters that in reality would have been too costly and time-consuming to evaluate. The Iowa Center for Emerging Manufacturing Technology at Iowa State University had built a fixed-base driving simulator that allowed users to test the feel of various automobile configurations. It relied on inputs attached to a steering wheel, an accelerator pedal and a brake. The sound of a small electric motor provided feedback to help the user "feel the road." The inputs were constantly updating the vehicle's position, speed and orientation. This provided car manufacturers with a cost-effective way to select the most comfortable designs and helped highway engineers design the safest road surfaces. We found virtual skid marks where an overconfident driver left the road.

Another goal of model building in the late 20th century was to represent objects too small for the naked eye to see. Computing technology was starting to enhance the understanding of very large or very small objects without requiring highly specialized devices like telescopes or electron microscopes and was being applied broadly in the fields of genetic engineering, chemistry, physics and just about all the other sciences. The *NanoManipulator,* developed at the University of North Carolina, actually enhanced a user's appreciation of these computerized representations. An input device let the user "feel" the intricate surface of a compact disk using a force-feedback input device. Its uses might have been the precise modification of extremely small physical objects.

Information Analysis and Synthesis

As information bases grew in size, it became increasingly difficult to pinpoint relationships and trends. *Starlight,* a combined project from Battelle Pacific Northwest Laboratories and the Boeing Company, allowed users to spotlight relationships through a series of graphical tools. Trends could be identified from a large collection of text or video input that would otherwise take weeks or months of analysis. The NASA Ames *Virtual Windtunnel* used virtual reality to let a user understand the mass of numbers produced during aerodynamic testing.

It then became possible to access databases through direct physical interaction. *Synthetic Interviews* from Carnegie-Mellon allowed a user to carry on an in-depth conversation with virtual figures as diverse as teachers, celebrities or politicians; or actor re-creations of historical figures such as Albert Einstein, Mark Twain or William Shakespeare.

The previous techniques were considered reactive in their approach because they required users themselves to find or analyze information. That's where Paleotechnics decided to let agents do the dirty work for them (without the standard 10 percent the Hollywood types were getting!). Work done by the Firefly Network developed collaborative filtering software, enabling companies to synthesize individual customers' tastes and preferences and gear products specifically to them. Agents would suggest certain movies and music, which saved users time and energy walking through the Internet.

Building Better Computer Systems

At the end of the millennium, the basic building blocks of computing technology were improving rapidly. The computer had finally evolved from a large package that could do a few simple functions to a small package that could do a lot of complex functions. Chip design, led by companies like Intel, was rapidly improving, allowing for the integration of high-quality graphics, video and audio. Distributed networked computers were rapidly replacing stand-alone computers, which led to the need for better circuitry. Japan's Aizu University built *Paradise-World,* an application that allowed VLSI circuits to be synthesized in minutes rather than hours. Paleotechnologists used programs like this to build faster computers faster.

In only a few years, computers had replaced mechanical devices as the primary control logic in numerous modern devices. Such real-time systems put additional stress on software-engineering methods to insure that control programs were highly reliable and stable. Korea's Pohang University of Science and Technology developed *ASADAL,* a computer-aided software-engineering environment designed to support requirements specification, analysis and validation for real-time systems. This cut down testing time in everything from cell phones to aircraft avionics.

Application software depends on the use of tools for improving the way in which software is developed. The Software Engineering Research Center, a consortium that included the University of Oregon, Ball State University, Purdue University and others, was in the process of developing a broad range of those tools. They included:

ProcSimity, for simulation and visualization of job-scheduling and processor-allocation algorithms for supercomputers like Intel Paragon, Cray T3E, and NCUBE.

Design Metrics Analysis, which helped identify error-prone software units and correct them during the design stage rather than during tests or actual production.

TAMER, which provided dependability assessments of distributed fault-tolerant systems. These assessments allowed users to identify the parts of a system most likely to fail.

Complex systems that rely on visualization for analysis or simulation were often built on components that simplified the graphics-intensive aspects of the problem. CGSD Corporation provided software packages that were used to develop tank simulators for the U.S. Department of Defense.

Interacting with the Computer

While the technology in the Paleotechnic Era was progressing at an extraordinary pace, the human side of the equation was evolving at a much slower rate. After all, from what we had pieced together about those who lived at the end of the 20th century, we found that one of their biggest technological complaints was the inability to set the clock on their VCR, let alone on a personal computer. Therefore, we concluded, the skittishness of those in the nascent information age to provide input and comprehend output was a major limiting factor in the computer's penetration into society.

Input Technology

The evolution of computing input techniques was slower than the evolution of computing systems themselves. While there was a substantial amount of variation in specific components, the earliest developers relied on computing behavior to be well-defined and predictable. By 1997, they finally began to understand that there was even less predictability when a person was introduced into the environment. Because of that, developers at the start of the information age were beginning to design input technology that made it much easier to input data and let the computer respond to the user, not vice versa.

The techniques being implemented were considered "natural inputs." LC Technologies, Stanford University and Twente University of the Netherlands were working on a variety of them. Eye-tracking techniques followed eye movements to control the computer; a company called BioControl used biofeedback and bioelectric-signal pattern recognition to facilitate user input directly from muscle tension and gestures. The techniques were used to help the physically disabled as well as to capture motion in virtual reality systems. They also helped monitor and avoid repetitive-stress injuries (RSI) that were common with keyboard-based systems.

Humans have the capacity to use multiple senses to acquire feedback as they work with computers. From that, new technology was developed that used feedback to prevent errors or guide the user through complex problems. So-called "haptic" feedback allowed for the development of a finger-driven harness or a joystick that let the user "feel" response (like hard or sticky) from an action. Developed by SensAble Technologies and Immersion, these devices allowed full movement in a 3D environment and were particularly effective in letting users feel collisions or the surface characteristics of objects. Cinematrix Interactive Entertainment Systems applied the feedback concept to multiple users, allowing a group of people to input information simultaneously and study the results together. This technology was particularly useful for gauging the opinions of an audience in real time and became popular at sales meetings and conferences like ACM97.

Output Technology

By 1997, it was becoming increasingly clear that pictures, rather than words and graphs, could summarize large amounts of data in a readily accessible, user-friendly form. Computer graphics, therefore, was an essential area of growth in the realm of computing technology.

Even the common screen as the primary viewing device was being expanded to take in a user's entire visual field. MicroVision was using optoelectronic technologies for patterning light to display and capture images. Its *Virtual Retinal Display* put information directly onto the human eye using a hand-held or head-worn device. It rapidly scanned a low-power colored laser onto the retina and created a high-resolution, full-motion image without screens or projection.

As the amount of information in a 3D picture increased, so did the need for greater performance. Most conventional graphics systems used the pixels on a raster screen to control the rendering process. The University of Mannheim in Germany constructed *VIRIM* to provide medical specialists with a way to better visualize tumors.

Holograms had typically used lasers to project 3D images into the air. Dimensional Media Associates created those images without the use of lasers though its *High Definition Volumetric Display* technology. The full-color, full-motion images appeared to float in space and could be seen from wide viewing angles under ambient lighting conditions. When this technology was first being developed, it seemed that people got a huge kick out of interacting with 3D holographic versions of historical figures.

3D Technology Labs carried the notion of laser projection one step further. Instead of using lasers to create an image in thin air, the lasers were projected into a specially treated glass cube. The result was a real three-dimensional image that an observer could walk around.

Cooperative Computing

After developing the technology to let users work individually or in small groups, the advent of high-speed networking technology and increased processing power at the end of the millennium made it possible for people to work together from anywhere in the world. Known as "distributed" computing, it let geographically dispersed people work together to construct complex environments in real time. As an educational tool, the Digital Circus developed project *CitySpace,* a virtual urban environment built collaboratively by kids, educators and media artists across the Internet. It invited young people to share stories, pictures, sounds and models of their own creation and assemble them into a navigable, 3D city model. Multiple participants could work simultaneously to build and modify the city. The Rochester Institute of Technology took the notion of building in another direction. Its *CAROL* project used the Web to build cooperation and collaboration across the far-flung Rochester arts community.

Cooperation in a distributed environment was applied to computers as well as people. The Ohio State University *Viento* project integrated two large-scale simulation models running on supercomputers in different locations. One in Colorado computed atmospheric conditions to simulate changing weather conditions. Another in Ohio used weather conditions to predict the evolution of physical lake conditions. Scientists in Michigan at a third site provided additional input data and used the two models to forecast changing weather patterns around Lake Erie. The San Diego Supercomputer Center worked to make a wide variety of high-performance computers conveniently available through the Web.

The Fraunhofer Center for Research in Computer Graphics (CRCG) showed how computing could allow people to literally work with one another simultaneously on difficult problems. We saw evidence that people in California, Rhode Island and Germany were able to concurrently develop a 3D model using now antiquated 1997 communications technology.

When users were working in multiple locations, it was important to know who was working with whom—and on what. The most reliable measure of what users were focusing their attention on was gaze direction. Twente University developed a prototype system called *GAZE,* an eye-tracking technique that allowed for multiparty conferencing and collaboration. The growth in eye-tracking technology allowed users to see who was talking to whom about what in a 3D meeting room on the Internet.

Conclusion

The late 20th century was clearly a period of great anticipation. With the growth of Internet-based applications, people everywhere were realizing that they could have access to any amount of information they wanted, any time they wanted it. Because of that, the world was growing smaller and, it seemed, smarter. Perhaps the most interesting thing about this era was the understanding that the evolution of computing technology, like the evolution of mankind itself, was a limitless process. Much of the technology we recovered from the San Jose site is in common use in 2047, and some has been replaced by better tools and techniques.

Having seen how far computer technology had come in the 50 years since EDSAC, ENIAC and the like, technologists in '97 had a bittersweet awareness: They clearly relished the fact that they were at the start of something great, but resigned themselves to the fact that they wouldn't be around to see it through. Little did they imagine how much would change in the next 50 years.

INDEX